STP 1466

Techniques in Thermal Analysis: Hyphenated Techniques, Thermal Analysis of the Surface, and Fast Rate Analysis

Wei-Ping Pan and Lawrence Judovits, editors

ASTM Stock Number: STP1466

ASTM
100 Barr Harbor Drive
PO Box C700
West Conshohocken, PA 19428-2959

Printed in the U.S.A.

Library of Congress Cataloging-in-Publication Data

Techniques in thermal analysis : hyphenated techniques, thermal analysis of the surface, and fast rate analysis / Wei-Ping Pan and Lawrence Judovits, editor.
   p. cm.
   "STP1466."
   ISBN 978-0-8031-5616-6
   1. Thermal analysis—Congresses. 2. Thermogravimetry—Congresses. I. Pan, Wei-Ping, 1954- II. Judovits, Lawrence, 1955-

   QD79.T38T384 2007
   543'.26—dc22   2007004395

Photocopy Rights

Authorization to photocopy items for internal, personal, or educational classroom use, or the internal, personal, or educational classroom use of specific clients, is granted by the American Society for Testing and Materials International (ASTM) provided that the appropriate fee is paid to the Copyright Clearance Center, 222 Rosewood Drive, Danvers, MA 01923; Tel: 978-750-8400; online: http://www.copyright.com/.

Peer Review Policy

Each paper published in this volume was evaluated by two peer reviewers and at least one editor. The authors addressed all of the reviewers' comments to the satisfaction of both the technical editor(s) and the ASTM International Committee on Publications.

The quality of the papers in this publication reflects not only the obvious efforts of the authors and the technical editor(s), but also the work of the peer reviewers. In keeping with long-standing publication practices, ASTM International maintains the anonymity of the peer reviewers. The ASTM International Committee on Publications acknowledges with appreciation their dedication and contribution of time and effort on behalf of ASTM International.

Printed in Mayfield, PA
August, 2007

Foreword

This publication, *Techniques in Thermal Analysis: Hyphenated Techniques, Thermal Analysis of the Surface, and Fast Rate Analysis,* contains papers presented at the symposium of the same name held at ASTM International Headquarters, W. Conshohocken, PA, on 24-25 May 2004, sponsored by the ASTM International Committee E37 on Thermal Measurements. The symposium chairmen were Prof. Wei-Ping Pan, Western Kentucky University, Bowling Green, KY and Dr. Lawrence Judovits, Arkema Inc., King of Prussia, PA.

Contents

Overview (updated 1/12/2007)

In May 2004 a two day symposium titled "Techniques in Thermal Analysis: Hyphenated Techniques, Thermal Analysis of the Surface, and Fast Rate Analysis" was held at the ASTM Headquarters in West Conshohocken, PA. Twenty-two presentations were given at the symposium. Additionally, the presenters were given the opportunity to submit to the Journal of ASTM International and for their papers to be included into a special technical publication (STP), thirteen papers were received.

The symposium itself was timely and reflected leading edge research in thermal analysis. Of major interest now is fast scan calorimetry in both instrument development and techniques. Through the use of a thin film nanocalorimeter scanning rates as high as 10,000 °C/sec can now be achieved. This, for example, allows for the better study of semicrystalline polymers where the reorganization process can be inhibited and the original metastable crystal can now be analyzed. Through the use of current technology, fast heating rates were employed to study epoxy curing. Fast rate analysis allowed the separation of the glass transition and cure exotherm.

The Hyphenated Techniques session brought some interesting papers mostly using thermogravimetric analysis (TGA) with another technique. It should also be noted that other techniques that have hyphens were also presented such as a paper on temperature-modulated differential scanning calorimetry, which is more prevalently written modulated temperature differential scanning calorimetry without the hyphen. An interesting study of the combined use of TGA with DTA (differential thermal analysis) and Raman spectroscopy was presented. The spectroscopy was performed on the sample itself as it underwent physical changes. This allowed the more precise study of dehydration of pharmaceuticals. Also presented was a paper advocating improved modeling when using hyphenated techniques such as TGA/FTIR (Fourier transform infrared) allowing kinetic parameters to be determined using both sets of data. Also of note was a simple calibration method for the quantitative use of mass spectrometry with TGA for a variety of encountered off gases.

Finally, a number of papers were given on thermal analysis of the surface. Many of these papers centered on the use of a modified atomic force microscope (AFM), or Micro-Thermal Analysis, that uses the AFM probe as a thermal device. A technique that shows promise is the use of micro-thermal analysis in combination with other techniques such as FTIR. This technique is referred to as photo thermal micro-spectroscopy (PTMS). PTMS uses the AFM probe to detect temperature fluctuations after a sample has been exposed to IR radiation allowing the construction of an infrared spectrum. This permits for a fast identification of an unknown material with minimal sample preparation.

The symposium chairs would like to acknowledge and extend our appreciation for all who have helped with the organization of the symposium and subsequent publications. A special thanks goes out to the reviewers who took the time and provided the needed commentary. Finally, we would like to recognize the sponsorship of both ASTM International Committee E37 on Thermal Measurements and the Thermal Analysis Forum of the Delaware Valley.

Prof. Wei-Ping Pan
Western Kentucky University
Bowling Green, Kentucky
Symposium Co-chairman and editor

Dr. Lawrence Judovits
Arkema Inc.
King of Prussia, Pennsylvania
Symposium Co-chairman and editor

HYPHENATED TECHNIQUES

Journal of ASTM International, Vol. 4, No. 1
Paper ID JAI100523
Available online at www.astm.org

Zhongxian Cheng,[1] *Hui-ling Chen,*[1] *Yan Zhang,*[1] *Pauline Hack,*[1] *and Wei-Ping Pan*[1]

An Application of Thermal Analysis to Household Waste

ABSTRACT: Thermal analysis combined with composition analysis has been used in this work to identify evolved gas when burning household waste. Thermogravimetry (TG) coupled with Fourier transform infrared (FITR) spectroscopy and mass spectrometry (TG-FTIR-MS) offers structural identification of compounds evolved during thermal processes. Carbon, hydrogen, and nitrogen elemental analysis and XRD offers raw materials information. All these combined can help to evaluate the chemical pathway for the degradation reactions by determining the decomposition products. For comparison purposes, emitted species concentrations are measured with multiple-sensors when burning household waste in a lab-scale fluidized bed combustor. There is excellent agreement between the two different approaches. The measured results also provided some insight into burning household waste as an energy source in large-scale incinerators.

KEYWORDS: thermal analysis, pyrolysis, combustion, household waste, TG-FTIR-MS

Introduction

Treatment of household waste is attracting a great deal of interest because of the increasing amount of household waste each year. Appropriate treatment of household waste can reduce the environmental pollution problem. In addition, waste-to-energy conversion could help to lower increasing energy costs. Many treatments have been used to deal with household waste, including landfilling and incineration [1]. Meanwhile, new techniques and methods are continuously being developed for waste treatment [2]. Among all the available techniques, incineration of household waste is a good solution because some energy could be recovered from the waste. Burning municipal solid waste (MSW) can generate energy while reducing the amount of waste by up to 90 % in volume and 75 % in weight [3]. The incineration process can generate steam, heat sources, and even electricity. However, it can also create pollutants or other hazardous material. The objective of this work is to investigate evolved species from general household waste using thermal analysis techniques. The result should help to further understand the mechanisms of the incineration process. Furthermore, it could help decision-makers select the proper approach for dealing with household waste according to regulations. More specifically, the emissions of hydrogen chloride (HCl), hydrogen sulfide (H_2S), and oxidized sulfur (SO_2) are the main concerns of this work.

Evolved gas analysis has been used extensively to identify qualitatively or quantitatively, or both, volatile products formed during the thermal degradation of materials. This technique involves the analysis of gaseous species evolved during combustion and pyrolysis in which a series of chemical reactions occur as a function of temperature and are analyzed using thermal analytical methods. Evolved gas analysis is normally used to evaluate the chemical pathway of degradation reactions by determining the composition of the decomposed products from various materials.

Simultaneous analysis of evolved gas with multiple instruments is a preferred method of detection for low concentrations of gas species. This method examines materials at the same time and could provide real-time results. One example of this analytical method is the combination of thermogravimetric, Fourier transform infrared spectroscopy, and mass spectroscopy analysis (TG/FTIR/MS), which is the primary method used for this work. Other combined analysis techniques that employ more than one sample for each instrument are used for household waste analysis, such as XRD for composition analysis.

Manuscript received March 3, 2006; accepted for publication October 4, 2006; published online November 2006. Presented at ASTM Symposium on Techniques on Thermal Analysis: Hyphenated Techniques, Thermal Analysis of the Surface and Fast Rate Analysis on 24 May 2004 in West Conshohocken, PA; L. Judovits and W.-P. Pan, Guest Editors.

[1] Western Kentucky University, Institute for Combustion Science and Environmental Technology, 2413 Nashville Road, Bowling Green, KY 42101.

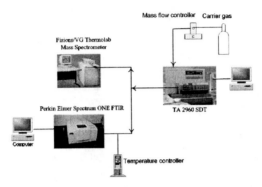

FIG. 1—*Schematic of TG-FTIR-MS setup.*

Experiment

Samples

Household waste is a highly heterogeneous mixture. The samples used in this work consist of combustibles waste (papers, newsprint, wood wastes, food waste) and noncombustible waste (plastic bags, coke-cans, glass, and other metals). All samples are predried and chopped into small pieces (less than 500 μm) and blended very well before analysis.

TG-FTIR-MS

A schematic of the unique coupled setup of TG-FTIR-MS is shown in Fig. 1. It consists of TA 2960 SDT,[2] Fisons/VG Thermolab Mass Spectrometer,[3] and PerkinElmer Spectrum ONE FTIR. The TA 2960 SDT is interfaced with a Fisons/VG Thermolab Mass Spectrometer by means of a heated capillary transfer line with 5 % of the evolved gas flowing into the MS system and the rest of the evolved gas flowing into the PerkinElmer FTIR[4] system. Experimental conditions are listed in Table 1. In order to simulate the natural conditions of the incinerator, two consecutive stages are studied. The first stage is a pyrolysis process. The sample is purged 20 min at the beginning and then it is heated from room temperature to 666 °C at a rate of 10 °C/min. The experiments take place in a nitrogen atmosphere with a flow rate of 100 mL/min. Then air is introduced and the second stage, the combustion process, begins. The rest of the sample is heated to 1270 °C at a rate of 10 °C/min. The capillary transfer line is heated to 120 °C, and the inlet port on the mass spectrometer is heated to 150 °C. The Fisons unit is based on a quadrupole design with a 1–300 amu mass range. The sample gas from the TGA is ionized at 70 eV. The system is operated at a pressure of 1×10^{-6} torr. Most of the evolved gas (95 %) flows into the FTIR gas cell. The purged gas flows at a rate of 100 mL/min and carries the evolved gases to a 70-mL sample cell with KBr windows via a silicone transfer line. The sample cell is wrapped with heat tape heated to 200°C to prevent the evolved gases from condensing. The sample cell is placed in the beam path of a PerkinElmer Spectrum ONE FTIR. The IR detection range is 4500 cm^{-1} to 500 cm^{-1}.

TABLE 1—*Experimental parameters.*

Isothermal Time (min)	Temperature Range in Nitrogen, (°C)	Heating Rate (°C/min)	Temperature Range in Air, (°C)	Heating Rate (°C/min)
20	20–666	10	667–1270	10

[2]TA Instrument, New Castle, DE.
[3]VG Gas Analysis Systems, Cheshire, England.
[4]PerkinElmer Ltd., Beaconsfield, United Kingdom.

TABLE 2—*Summary of TGA data.*

Weight Loss in Pyrolysis Stage (%)	Weight Loss in Combustion Stage (%)	Residue (%)
60.7	13.0	30.3

Composition Analysis of Raw Sample

Carbon (C), hydrogen, and nitrogen analysis is conducted using a LECO CHN 2000 Analyzer in accordance with ASTM Standard Test Methods for Instrumental Determination of Carbon, Hydrogen, and Nitrogen in Laboratory Samples of Coal and Coke (D 5373). Moisture, ash, and volatile matter analysis is conducted using a LECO TGA 601 System according to ASTM Standard Test Methods for Proximate Analysis of the Analysis Sample of Coal and Coke by Instrumental Procedures (D 5142). Sulfur analysis is conducted using a LECO SC432 Sulfur Analyzer according to ASTM Standard Test Methods for Sulfur in the Analysis Sample of Coal and Coke Using High-Temperature Tube Furnace Combustion Methods (D 4239). Mercury analysis is conducted using a LECO AMA 254 Mercury Analyzer according to ASTM Test Method for Total Mercury in Coal and Coal Combustion Residues by Direct Combustion Analysis (D 6722). BTU analysis is conducted using a LECO AC350 Bomb Calorimeter according to ASTM Standard Test Method for Gross Calorific Value of Coal and Coke (D 5865). Chlorine analysis is conducted using a LECO AC350 Bomb Calorimeter and Dionex Ion Chromatograph according to ASTM Standard Test Method for Total Chlorine in Coal by the Oxygen Bomb Combustion/Ion Selective (D 4208) (in this work, the Ion Chromatograph instrument is substituted for the ion selective electrode). Major and minor elements are analyzed using a Rigaku RIX 3001 XRF analyzer according to ASTM Standard Test Method for Major and Minor Elements in Coal and Coke Ash by X-Ray Fluorescence (D 4326).

Results and Discussions

Thermogravimetric Data Analysis

Thermogravimetric analysis can provide information on kinetics and on reactivity of waste material [4]. It is expected that the thermal conversion in the incineration process goes through two different stages: the pyrolysis of volatile material at a low temperature range and the combustion of nonvolatile material at a high temperature range. The weight loss and main kinetics are shown in Table 2 and Fig. 2. The majority (60.7 %) of the sample is volatile material and evolved gases come out during the pyrolysis stage (25 to 666 °C). Also, it is found that there are two main peaks that indicate reactions in this stage, occurring at 337 °C and 420 °C, respectively (see Fig. 2). The possible evolved gases for the first peak (337 °C) include H_2O, CH_4, SO_2, HCl, and H_2S (see FTIR plot in Fig. 7.). There is a 25 % weight loss after the first main peak. Most of these species come from pyrolysis of wastes, such as plastics. The second peak at 420 °C shows that more volatile gases are released and another 35.5 % weight loss occurs after this peak. The possible evolved gases for this are H_2O, CH_4, C_2H_4, SO_2, HCl, and H_2S. At the end of pyrolysis, air is introduced into the furnace instead of nitrogen. Combustion occurs and a third peak is

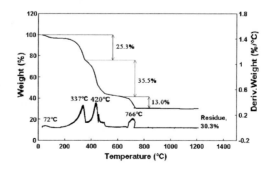

FIG. 2—*TGA curves for pyrolysis stage and combustion stage.*

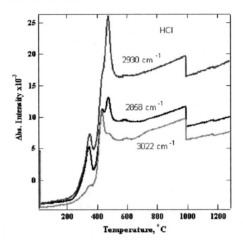

FIG. 3—*FTIR for HCl.*

located at 766 °C. Nonvolatile materials are burnt at this high temperature and a 13.0 % weight loss accompanied this process. The primary emitted gas is carbon dioxide (CO_2), a combustion product, as verified in Fig. 6 and Fig. 9.

FTIR Data

Fourier transform infrared spectroscopy (FTIR) can provide some identification information for the complicated pyrolysis process when combined with other techniques [5]. Regulations for species emission from incineration set the limits. So evolved gases (such as HCl, Cl_2, H_2S, SO_2) and their emitted level are the main concerns for this work. The individual species spectra are shown in Figs 3–6. The FTIR spectra of emitted HCl at wave numbers of 2930 cm^{-1}, 2858 cm^{-1}, and 3022 cm^{-1} are shown in Fig. 3. HCl gas initially emits at around 250 °C and comes to peak emission at around 450 °C. The FTIR spectra of emitted SO_2 (1375 cm^{-1} and 1344 cm^{-1}) are shown in Fig. 4. Much like HCl, SO_2 (Fig. 5) initially emits at around 220 °C and has a peak emission at around 400 °C. H_2S (1180 cm^{-1}, 1283 cm^{-1}, and 2173 cm^{-1}) seems to initially emit slightly later at around 300°C, and it reaches its peak emission at 400 °C. The

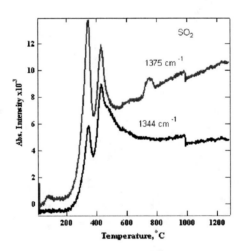

FIG. 4—*FTIR for SO$_2$.*

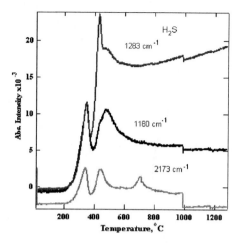

FIG. 5—*FTIR for H₂S.*

FTIR spectra of CO_2 (669 cm^{-1} and 2357 cm^{-1}) are shown in Fig. 6. Obviously CO_2 emissions mainly occur at around 800 °C, when combustion dominates the thermal conversion process.

In addition, FTIR spectra at fixed temperature scans are shown in Figs. 7–9. Figure 7 shows the emitted species spectra at 330 °C during the pyrolysis stage. The typical emitted gases of this stage are HCl, H_2S, H_2O, and CO_2 as mentioned earlier. Figure 8 shows the emitted species spectra at 450 °C during the pyrolysis stage with the emitted gases being C_2H_4, HCl, H_2S, H_2O, and CO_2. The spectra of emitted gases at 730 °C (combustion stage) are shown in Fig. 9. CO_2 is the primary species emission during this stage.

MS Data

Mass spectroscopy (MS) also provides additional important identification information which the FTIR may not provide. For example, infrared spectroscopy is not available for some diatomic molecules that do not have a permanent dipole moment. Some important mass spectra during the pyrolysis/combustion process are shown in Fig. 10. However, MS measures only mass-to-charge ratios and has limits when determining isomers. The mass-to-charge ratios of (m/e=16,17,18) are listed in the same plot (Fig.

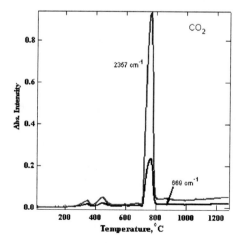

FIG. 6—*FTIR for CO₂.*

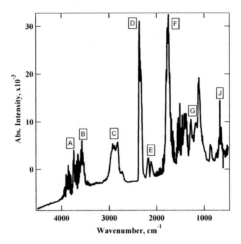

FIG. 7—*FTIR spectra at 330 °C (A: 3738 cm^{-1}, 3595 cm^{-1}; B: 3574 cm^{-1}; C: 3022 cm^{-1}, 3930 cm^{-1}, 2858 cm^{-1}; D: 2357 cm^{-1}; E: 2173 cm^{-1}, 2111 cm^{-1}; F: 1702 cm^{-1}; G: 1283 cm^{-1}; J: 669 cm^{-1}.*

10(*a*)). It indicates possible emissions of CH_4, O, or NH_3. The final identification of those species can be determined based on the combined consideration of FTIR spectroscopy and even knowledge of the sample itself. One example of this limit of MS measurement is the m/e=34 shown in Fig. 10(*b*). Based on mass-to-charge ratio, it may be H_2S (m/e=34) at 800 °C. But it conflicts with the FTIR finding shown in Fig. 5 which indicates that the peak of FTIR should be at around 400 °C. Therefore the determination of m/e=34 needs to incorporate other techniques. CO_2 (m/e=44) mass spectroscopy, shown in Fig. 10(*c*), agrees very well with the FTIR spectroscopy shown in Fig. 6. MS spectra of benzaldehyde (m/e=77) are shown in Fig. 10(*d*).

A data summary from both the FTIR and the MS is listed in Table 3. The detected species, such as CH_4, NH_3, H_2S, and HCl, their mass-to-charge ratio, their wavenumber, and their evolution temperature are given in this summary.

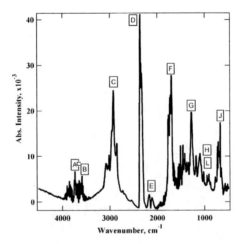

FIG. 8—*FTIR spectra at 450 °C (A: 3738 cm^{-1}, 3595 cm^{-1}; B: 3574 cm^{-1}; C: 3022 cm^{-1}, 3930 cm^{-1}, 2858 cm^{-1}; D: 2357 cm^{-1}; E: 2173 cm^{-1}, 2111 cm^{-1}; F: 1702 cm^{-1}; G: 1283 cm^{-1}; J: 669 cm^{-1} H: 1180 cm^{-1} I: 948 cm^{-1} J: 669 cm^{-1}.*

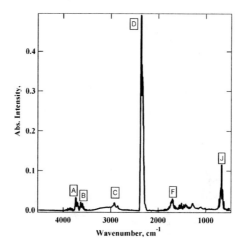

FIG. 9—*FTIR spectra at 730 °C (A: 3738 cm^{-1}, 3595 cm^{-1}; B: 3574 cm^{-1}; C: 3022 cm^{-1}, 3930 cm^{-1}, 2858 cm^{-1}; D: 2357 cm^{-1}; F: 1702 cm^{-1}, G: 1283 cm^{-1}; J: 669 cm^{-1}.*

Analysis of Raw Sample

Proximate analysis, ultimate analysis, and miscellaneous analysis for samples according to related ASTM methods are listed in Table 4. The samples are predried; therefore moisture is relatively low. The carbon content is 48 % (dry based). The measured calorific value is around 21 200 kJ/kg that could be recovered for further use. In addition, major compound measurements by XRF are also listed in Table 5.

Concentration Measurement of Emitted Gas from Combustor

In order to measure the emitted gas concentration, a large amount of the sample is burnt in a lab-scale fluidized bed combustor (see Fig. 11). This combustor is made of stainless steel pipe with a 50-mm inner diameter and a 700-mm effective length. The compressed air can be supplied into this reactor from three locations: through the top with sampling feeding, through the middle of the reactor, and through the bottom of the wind box and gas distributor. The flue gas exits from the top side to a gas analyzer by passing through a small cyclone. The IMR 5000 gas analyzer is used to analyze the flue gas composition. Twelve sensors are used simultaneously to measure species concentrations; there are sensors for O_2, CH_4, NH_3, H_2S, HCl, CO_2, CO, SO_2, CO_2, Cl_2, NO, and NO_2. The measured results are listed in Table 6.

Based on the TG-FTIR-MS analysis of burning waste, species of interest, H_2S, HCl, NH_3, and CH_4 are detected during pyrolysis and combustion. These species are further detected from the lab-scale combustor. A comparison of the findings given in Table 3 and Table 6 shows excellent agreement between the TG-FTIR-MS measurement and the multiple sensor-based measurements. Moreover, the TG-FTIR-MS measurement provides species emission history as temperature changes, while sensor-based measurements only measure total species concentration. The information from the TG-FTIR-MS measurement will allow further kinetic and mechanisms study for the pyrolysis and combustion process.

Estimated Maximum (Worst Case) Emission Level

In order to determine the level of emitted gas from combustion, estimation is performed based on theoretical calculation by assuming all species in fuel (household waste) are transferred to flue gas. This assumption will predict the worst case and maximum emission level. Table 7 lists the estimated maximum emission level. The allowable emission limit is also listed for comparison purposes. From Table 7, NO, SO_2, and HCl emissions based on the worst-case calculation will exceed the set-limit, but mercury emission even in the worst-case calculation is lower than the limit. The worst-case is seldom the actual finding because it is impossible that all species will be transferred into the flue gas. Fly ash and residue can hold some of them.

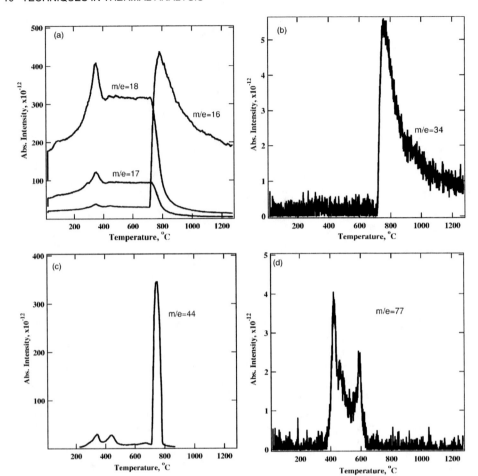

FIG. 10—*MS spectra: a: m/e = 16, 17, and 18; b: m/e = 34; c: m/e = 44; d: m/e = 77.*

TABLE 3—*Summarized results for MS and FTIR testing.*

m/z (MS)	Wavenumber [6] (FTIR) (cm⁻¹)	Occurred in Pyrolysis Stage (°C)	Occurred in Combustion Stage (°C)	Possible Evolved Gas
1		365		H
16	3022	104	753	CH₄
17	932	222		NH₃
18	3738, 3595	162	698	H₂O
26	732	228	698	C₂H₂
28	948	262		C₂H₄
32	1035	202		CH₃OH
34	1180, 1283, 2173	222		H₂S
36	2930, 3022, 2858	246		HCl
44	2357, 669	222	698	CO₂
45	679	222		C₂H₅O
64	1344, 1375	123		SO₂
77, 78	689	242	698	benzaldehyde
	1733, 1761	192	698	ketone

TABLE 4—*Proximate, ultimate, and miscellaneous analyses.*

	As Received Basis	As Determined Basis	Dry Basis	Dry, Ash-free Basis
Air dry loss moisture	0.0			
Proximate Analysis				
Moisture, %	5.3	5.3		
Ash, %	19.90	19.90	21.01	
Volatile, %	69.11	69.11	72.98	92.39
Fixed carbon, %	5.69	5.69	6.01	7.61
Ultimate Analysis				
Carbon, %	46.25	46.25	48.84	61.83
Hydrogen, %	7.33	7.33	7.12	9.01
Nitrogen, %	0.37	0.37	0.39	0.50
Sulfur, %	0.40	0.40	0.42	0.53
Oxygen, %	25.16	25.16	21.60	27.34
Miscellaneous analysis				
Calorific value, kJ/kg	20 075	20 075	21 200	26 837
Chlorine, ppm	5881	5881	6210	7862
Fluorine, ppm	73	73	77	98
Mercury, ppm	0.21	0.21	0.22	0.28

TABLE 5—*Major compound analyses by XRF.*

Na_2O (%)	MgO (%)	Al_2O_3 (%)	SiO_2 (%)	P_2O_5 (%)	SO_3 (%)	K_2O (%)	CaO (%)	TiO_2 (%)	MnO (%)	Fe_2O_3 (%)	BaO (%)	SrO (%)
6.12	1.86	11.73	44.86	1.18	3.29	1.72	16.76	2.18	0.16	2.07	0.14	0.07

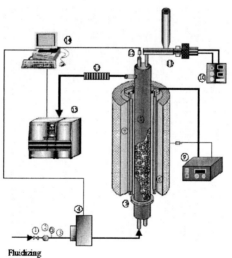

Fluidizing Gas

1. Valve	6. Electric Heater	11. Coal Feeder
2. Regulator	7. Insulation	12. Thermocouple
3. Gage	8. Fluidized Bed	13. Flue Gas Cooler
4. MFC	9. Temperature Controller	14. Computer
5. Windbox	10. Feeder Controller	15. Flue Gas Analyzer

FIG. 11—*Lab-scale fluidized bed combustor.*

TABLE 6—*Flue gas concentration when household waste is burned in a lab-scale fluidized bed combustor.*

	Condition A[a]	Condition B[b]	Condition C[c]
O_2, %	5.5–6.0	6.1–6.8	5.3–5.8
CO_2, %	13–14.5	3.3–6.8	10.1–11.7
CH_4, %	0	0	0
CO, ppm	3900–4300	3970–4270	3960–4285
SO_2, ppm	250–350	133–150	450
NH_3, ppm	180–280	99–125	205–398
H_2S, ppm	150–180	50–70	166–330
HCl, ppm	200–300	90–154	165–180
Cl_2, ppm	10–20	8–11	19–29
NO, ppm	50–150	99–150	168–252
NO_2, ppm	0	0	0

[a]Air flow was supplied from the top location.
[b]Air flow was supplied from the top and bottom location.
[c]Air flow was supplied from the top and middle location.

TABLE 7—*Estimated maximum emission level based on theoretical calculation.*

	Calculation	Allowable Emission Limit
NO, ppm	1048	38
SO_2, ppm	495	20
HCl, ppm	600	62
Hg, mg/dscm	0.037	0.47

Conclusions

Evolved gas analysis with the TG-FTIR-MS and related element/compound analysis are conducted for the pyrolysis and combustion process of typical household waste. There is about a 60 % weight loss and a large amount of evolved gas emission during pyrolysis. These gases include HCl, H_2S, NH_3, and SO_2. The absolute concentrations for these evolved gases are not determined from the TG-FTIR-MS measurement, but the relative level can be estimated from the elemental analysis of the raw sample. The TG-FTIR-MS measurement can provide important information for further kinetics study of household waste burning during the pyrolysis/combustion process. In addition, emitted species concentrations are measured with multiple-sensors when household waste is burning in a lab-scale fluidized combustor. There is excellent agreement between the TG-FTIR-MS measure and the multiple-sensors measurement. The estimated maximum emission level is calculated based on the assumption that all species are transferred into flue gas. It is also found that there is around a 20 000 kJ/kg calorific value for the partial energy of household waste that may be recovered if appropriate treatment is applied.

References

[1] Wenisch, S., Rousseaus, P., and Metivier-Pignon, H., "Analysis of Technical and Environmental Parameters for Waste-to-Energy and Recycling: Household Waste Case Study," *Int. J. Therm. Sci.*, Vol. 43, 2004, pp. 519–529.

[2] Zhang, Y., Deng, N., Ling, J., and Xu, C., "A New Pyrolysis Technology and Equipment for Treatment of Municipal Household Garbage and Hospital Waste," *Renewable Energy*, Vol. 28, 2003, pp. 2383–2393.

[3] Solid Waste Combustion and Incineration, http://www.epa.gov/eaposwer/non-hw/muncpl/landfill/sw_combst.htm

[4] Agrawal, R. K., "Compositional Analysis of Solid Waste and Refuse Derived Fuels by Thermogravimetry," *Compositional Analysis by Thermogravimetry*, C. M. Earnest, Ed., ASTM International, West Conshohocken, PA, 1988, pp. 259–271.

[5] Xie, W., and Pan, W.-P., "Thermal Characterization of Materials Using Evolved Gas Analysis," *J. Therm Anal. Calorim.*, Vol. 65, 2001, pp. 669–685.

[6] Hanst, P. L., and Hanst, S. T. (Infrared Analysis, Inc), "Infrared Spectra for Quantitative Analysis of Gases," in *Gas Analysis Manual for Analytical Chemists in Two Volumes*.

Journal of ASTM International, October 2005, Vol. 2, No. 9
Paper ID JAI12789
Available online at www.astm.org

Wendy J. Collins,[1] Corey DuBois,[1] R. Thomas Cambron, Ph.D.,[2] Nancy L. Redman-Furey, Ph.D.,[3] and Adrienne S. Bigalow Kern[4]

Development and Evaluation of a TG/DTA/Raman System

ABSTRACT: Modification of a commercially available TG/DTA instrument enabled development of a working TG/DTA/Raman system. The TG/DTA was modified by replacing a portion of the furnace wall above the sample and reference holders with a double layer insulated quartz window. The quartz window provided ready access for Raman spectroscopy, enabling simultaneous collection of TGA, DTA, and Raman data during thermal analytical experiments. Insertion of the quartz glass window did not impact TGA or DTA performance as demonstrated by the comparison of baseline drift and temperature calibration for modified and unmodified furnaces.

KEYWORDS: Thermal Analysis, TGA, TG/DTA, Raman, Raman Spectroscopy

Introduction

TG/DTA is typically used as a general survey tool. Mass losses observed in the thermal profile are not always easily interpreted. To aid in the interpretation, experiments have historically been conducted by combining the TG/DTA with other instrumentation that can analyze the evolved gases during the experiment [1–3]. The interpretation of results from the evolved gasses can aid in the identification of volatile degradation products [2,3]. Alternatively, samples have been isolated during the experiment and analyzed off-line by other relevant analytical techniques [3,4]. In addition to interrupting the thermal experiment, changes in the sample during preparation for analysis limit the potential of this approach.

Feasibility has been demonstrated for exploiting Raman spectroscopy to directly monitor changes in sample composition and phase transformations as a function of temperature [5–12]. These experiments have been performed by the use of a temperature controlled oven that mimics a TGA and allows collection of spectroscopic data concurrently with thermal analysis. This technique has been termed "thermo-Raman spectroscopy" by the authors. Thermo-Raman spectroscopy has also been coupled directly to a commercial TGA [13–15]. This coupled measurement offers advantages over thermo-Raman spectroscopy by ensuring consistent experimental sampling conditions for each individual measurement as well as the ability to observe the sample directly (solid state) without interruption during the thermal experiment. Experiments summarized in this report use this approach with the addition of the DTA to provide added insight into the thermal profile. A Raman spectrometer was coupled directly to a commercial TG/DTA with modifications to the existing furnace. Modifications required for conducting these measurements and the head-to-head comparison of the traditional and quartz-

Manuscript received 24 May 2004; accepted for publication 26 January 2005; published October 2005. Presented at ASTM Symposium on Techniques in Thermal Analysis: Hyphenated Techniques Thermal Analysis of the Surface and Fast Rate A on 24-25 May 2004 in West Conshohocken, PA.
[1] Senior Researcher, Procter & Gamble Pharmaceuticals, P.O. Box 191, State Route 320, Norwich, NY 13815.
[2] Senior Scientist, Procter & Gamble Pharmaceuticals, P.O. Box 191, State Route 320, Norwich, NY 13815.
[3] Principle Scientist, Procter & Gamble Pharmaceuticals, P.O. Box 191, State Route 320, Norwich, NY 13815.
[4] Senior Researcher, Procter & Gamble Pharmaceuticals, P.O. Box 191, State Route 320, Norwich, NY 13815.

windowed furnaces are presented in this report. Several compounds were selected based on their Raman scattering properties to assess the impact of laser illumination on the TG/DTA system. The samples, as shown in Fig. 1, included: naphthalene, microcrystalline cellulose, lactose monohydrate, and carbomer.

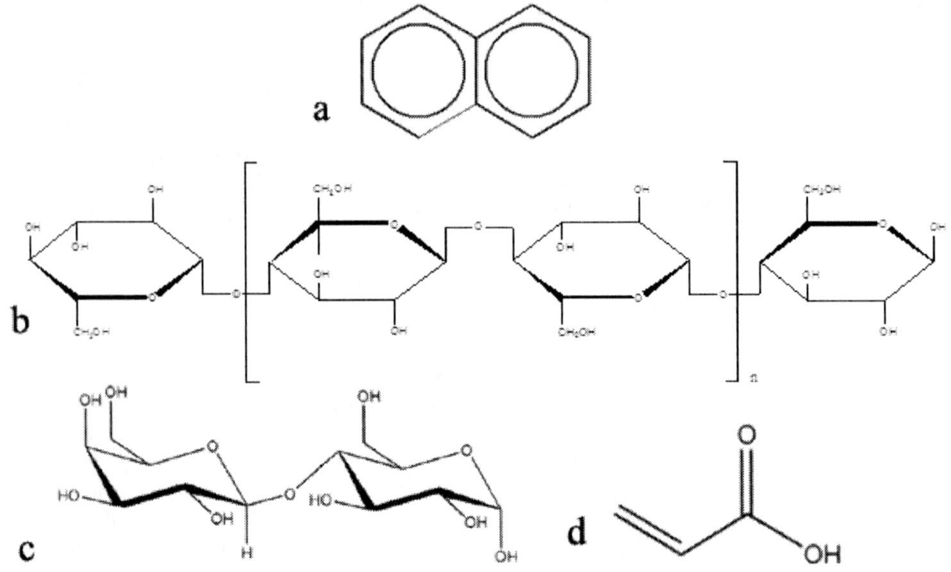

FIG. 1—*a) naphthalene, b) microcrystalline cellulose, c) lactose monohydrate, and d) carbomer.*

Experimental

Modification of a TG/DTA Furnace Assembly

A commercially available Seiko TG/DTA 220 was modified to allow the Raman probe unobstructed access to the sample. The modification consisted of imbedding a quartz window within the ceramic core of an existing furnace. A secondary quartz sleeve was needed to maintain the same heating effect as a traditional furnace assembly. The operational range was reduced as a result of these modifications. The range of the traditional furnace is from room temperature to 1050°C. As a result of imbedding the quartz window, the modified furnace's upper range was reduced to 550°C. A picture of a traditional and modified furnace assembly is shown in Fig. 2.

FIG. 2—*Traditional TG/DTA furnace on the right, modified furnace with quartz window shown on the left.*

Raman Spectroscopy

Raman spectra collected in these experiments utilized a Kaiser Raman Holoprobe spectrometer (785 nm laser excitation, 400 mW) coupled to a filtered fiber optic probe head. A non-contact collection optic with a 2.5-in. focal distance allowed laser illumination of the sample. This configuration enabled collection of spectra through the imbedded quartz window. The Raman probe head was positioned using multiple translational stages enabling alignment of the laser beam with the sample. A photograph of the coupled instrument is shown in Fig. 3.

FIG. 3—*TG/DTA/Raman assembly showing the translational stages used to position the Raman laser.*

Suitable Raman acquisition exposure intervals depend on the Raman scattering cross-section of each sample being investigated. Raman scattering cross-sections for the samples studied in these experiments are not known. Suitable Raman exposure intervals are determined based on signal-to-noise characteristics of the resulting spectrum. Samples with small scattering cross-sections require longer exposure intervals to produce a spectrum with suitable signal to noise. Conversely, samples with large scattering cross-sections require shorter exposure intervals to produce a suitable spectrum. Collection times used for these experiments ranged from 10–30 s.

TG/DTA/Raman Experimental Conditions

To successfully obtain quality Raman and thermal data, the position of the Raman probe-head relative to the sample was optimized prior to each experiment. The Raman laser was centered on the sample, and the focus of the Raman collection optic was optimized in a darkened room using multiple translational stages. Once the focus was optimized, the appropriate Raman acquisition exposure time was determined for each sample based on a suitable signal to noise ratio. The data collected from each spectrum were then subjected to a cosmic ray filter. With the position of the Raman probe-head optimized, the TG/DTA was loaded with a fresh sample, and the TG/DTA run was started simultaneously with the Raman collection.

Results and Discussion

Confirmation of Furnace Stability

Verifying that the baseline and calibration of the modified furnace were equivalent to the traditional furnace was the first objective. The baseline scans were measured in μg of drift from room temperature to 375°C. The head-to-head comparison yielded the following results: a total drift of 1.9 μg from 25–375°C for the traditional furnace, versus a total drift of 0.9 μg for the modified quartz furnace over the same temperature range. The insertion of the quartz window did not impact baseline drift of the TGA signal. The temperature calibration was verified by comparing the melt temperatures of recognized standards to accepted values. Table 1 shows the comparison of the traditional to the modified furnace with respect to theory for the melt temperatures of gallium and tin as defined by the International Temperature Scale of 1990. As shown in the table, the temperature calibration of the modified furnace could not be distinguished from that of the traditional furnace.

TABLE 1—*Melt temperature calibration data.*

	Gallium Theory 29.8°C	Tin Theory 231.9°C
Traditional Furnace (calibration 1 / calibration 2)	30.6°C / 28.7°C	231.7°C / 231.9°C
Modified Quartz Furnace (calibration 1)	29.2°C	232.1°C

Optimization of the TG/DTA/Raman System to Maintain Thermal and Spectral Quality

An important goal of this study was to obtain spectral information by Raman spectroscopy without compromising the spectral or thermal response of the sample. Several compounds were selected based on their Raman scattering properties to assess the impact of the laser upon the TG/DTA system. The samples included in this study were: naphthalene, microcrystalline cellulose, lactose monohydrate, and carbomer. The majority of the experiments were run using naphthalene, which is a strong Raman scatterer, and a recognized standard.

In an initial experiment, naphthalene was scanned from 25–100°C at 1°C/min, while Raman data were collected with the laser set at full power (400 mW), collection at 10-s intervals with a 10-s cosmic ray filter. As shown in Fig. 4, this arrangement produced suitable spectra with adequate signal to noise (Fig. 4c), but the laser output interfered with the thermal response (Fig. 4b). The corresponding thermal profile obtained without concurrent Raman collection is shown in Fig. 4a. The TGA trace in the absence of the Raman laser showed a gradual loss from the start of the scan to the end due to sublimation of naphthalene and a small artifact due to sample movement while the sample melted. The DTA baseline was stable up to the onset of melting of naphthalene with the baseline returning to normal after the melt.

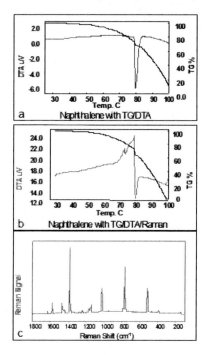

FIG. 4—*(Exotherm up) At 1°C/min with full laser power, excellent Raman spectra are collected, but the TG/DTA exhibits an obvious artifact due to laser heating of the sample:* a) *Naphthalene without Raman laser on;* b) *Naphthalene with concurrent Raman collection;* c) *Raman spectrum obtained during TG/DTA scan.*

When the experiment was repeated with the Raman laser energized, naphthalene spectra with adequate signal to noise were obtained from the start of the experiment until the loss of signal just prior to the melt. Spectra were acquired at 1-min intervals until the signal was lost.

However, the thermal profile as shown in Fig. 4b showed apparent heating of the sample by the laser prior to the melt onset. A second experiment was designed to determine if a faster scan rate (5°C/min) would result in less laser artifact because of a shorter period of exposure to the laser. Keeping all other conditions the same, the experiment was repeated at 5°C/min (Fig. 5). The Raman produced quality spectra with signal to noise comparable to that collected at 1°C/min (Fig. 5c). The DTA response improved significantly, showing a less obvious heating artifact prior to the melt. However, the baseline mismatch before and after the melt indicated that the Raman collection did interfere with the thermal response, although to a lesser degree than previously observed.

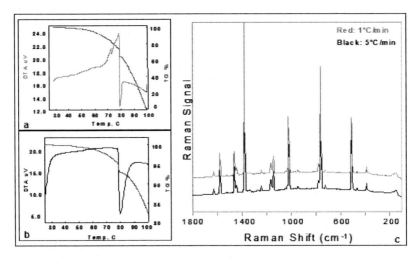

FIG. 5—(Exotherm up) Minimizing the TG/DTA artifact by running the concurrent analysis at 5°C/min instead of 1°C/min: a) Naphthalene with Raman laser on at 1°C/min; b) Naphthalene with Raman laser on at 5°C/min; c) Raman spectrum overlay of 1 and 5°C/min thermal scan.

The next step in obtaining an artifact free thermal response while collecting Raman data was to evaluate the Raman parameters that could be adjusted to minimize impact upon the thermal profile. The two parameters that were assessed were laser intensity and laser focus. Since the analysis at 5°C/min produced less alteration of the thermal profile, the next experiments were run at 5°C/min. Initially, the intensity of the laser was at full power or 400 mW output (60 mW at the sample). The power was initially reduced by 50 %, which resulted in a cleaner thermal profile, but still not optimal. A further reduction of 25 %, to a total reduction of 75 % of full power, produced an acceptable quality spectrum with less intensity but still clearly distinguishable as naphthalene (Fig. 6). An ASTM reference peak at 1382.2 cm⁻¹ was checked for each of the three laser output levels with some indication of linearity. The DTA quality improved dramatically as shown by comparison of 5°C/min scans with and without Raman collection (Fig. 7). Use of a 5°C/min scan rate with the laser output set at 25 % of full power produced acceptable Raman results and changed the TG/DTA curves minimally. No obvious difference was observed in the TGA curve, and shape of the DTA was maintained. The only evidence of Raman interference was the reduction of the melt area by approximately 20 %.

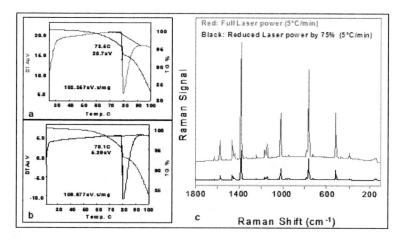

FIG. 6—*(Exotherm up) Reduction of thermal artifact by reducing the Raman laser power: a) TG/DTA with the Raman at full power; b) TG/DTA with the Raman at 25 % power; c) Raman spectra overlay of full and 25 % power.*

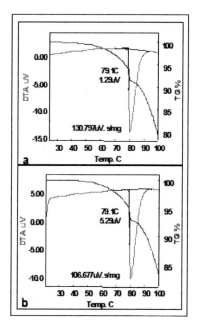

FIG. 7—*(Exotherm up) Comparison of 5°C/min scans of naphthalene with Raman collection at 25 % laser power and without Raman collection. At reduced laser power, little evidence of laser effect is seen upon the TG/DTA curve: a) TG/DTA without Raman collection; b) TG/DTA with the Raman at 25 % power.*

An experiment was then performed to determine the effect of laser focus on the DTA artifact (Fig. 8). In this experiment, naphthalene was placed in a sample pan and left inside the TG/DTA with the instrument at room temperature. The Raman laser was defocused both toward and away from the sample. Holding the TG/DTA isothermal at room temperature, the laser was left on until a steady DTA deflection was recorded. In this manner the magnitude of laser-induced heating could now be determined quickly without having to run a complete thermal scan. The initial experiment with naphthalene, shown in Fig. 8a shows the DTA response with the Raman laser at full power. As noted previously, good quality Raman spectra were obtained with this TG/DTA/Raman configuration, but it produced an unacceptable thermal profile. Figure 8b with the Raman focused 1 cm nearer to the sample shows a slightly reduced DTA response. Figure 8c with the Raman optic moved as close to the sample as possible shows a DTA response slightly greater than that from initial (Figure 8a). Figure 8d with the Raman optic moved as far from the sample as possible showed the greatest reduction with defocusing, however at this distance from the sample a Raman spectrum was not obtainable. By comparison, Fig. 8e shows the DTA response obtained by reducing the laser output by 75 %, the condition that resulted in the most acceptable TG/DTA curve. The DTA response here was less than one fourth of the response obtained at full power (Figure 8a) and significantly reduced compared to the best reduction observed by defocusing. This set of experiments demonstrated that laser focus was not a useful parameter in reducing artifacts caused by laser heating as was laser power.

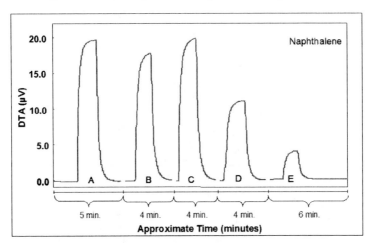

FIG. 8—*(Exotherm up) DTA response with respect to laser output and focus: a) the Raman at 100 % power, sample in focus; b) Raman focused away from the sample by 1 cm; c) the Raman optic moved as close to the sample as possible; d) the Raman optic moved as far from the sample as possible, spectrum not obtainable; e) the Raman laser output reduced by 75 %.*

The DTA response observed during controlled exposure from the laser experiment (isothermal TG/DTA, the Raman laser was cycled off and on) did not correspond solely to laser output. The DTA response appeared to be a function of both laser output and the spectral properties of the sample. Figure 9 shows the DTA response at 100 % laser power and the corresponding Raman spectra for the four compounds examined in this study. Naphthalene and lactose monohydrate display a large Raman scattering cross-section relative to microcrystalline cellulose and carbomer. Generally the samples with the large scattering cross-sections also had a larger DTA artifact; this does not appear to be the norm for all samples tested, indicating that other non-radiative processes subsequent to laser exposure may be involved.

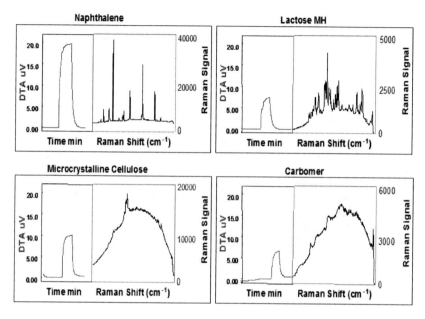

FIG. 9—*(Exotherm up) DTA response artifacts with associated Raman spectrum of naphthalene, lactose monohydrate, microcrystalline cellulose, and Carbomer.*

Summary

Insertion of a quartz window into a commercial TG/DTA produced a system that provided access to real time Raman monitoring of TG/DTA samples without degrading the furnace baseline or temperature calibrations. The DTA signal was found to be more sensitive to artifacts caused by heating due to the Raman laser than was the TGA. Increasing the TG/DTA scan rate (to reduce laser exposure by reducing the length of the run) and decreasing the laser output energy resulted in a thermal profile of naphthalene with minimal artifact and acceptable quality Raman spectra. The magnitude of DTA artifact induced by laser heating was shown to be affected by a combination of the laser output energy and the interaction of the laser with the sample. The fundamental principles of this interaction were not investigated in these experiments. However, these initial results indicate this hyphenated analytical methodology, in addition to providing spectral information to assist in the interpretation of thermoanalytical data, may be useful for providing calorimetric data from interactions between laser light and the materials under investigation.

Acknowledgments

The authors thank RT Instruments, Inc. of Woodland, California for the design modification and Roberts Instrument & Tool for the customization of the quartz optical furnace.

References

[1] Szekely, G., et al., *Thermochimica Acta*, Vol. 196, 1992, pp. 511–532.
[2] Kinoshita, R., et al., *Thermochimica Acta*, Vol. 222, 1993, pp. 45–52.
[3] Giron, D., *Journal of Thermal Analysis and Calorimetry*, Vol. 68, 2002, pp. 335–357.
[4] Vora, A., et al., *29th NATAS Conference Proceedings*, 2001, pp. 101–105.
[5] Chang, H., et al., *Reviews in Analytical Chemistry*, Vol. 20, No 3, 2001, pp. 207–238.
[6] Chang. H., et al., *Analytical Chemistry*, Vol. 69, 1997, pp. 1485–1491.
[7] Huang, P., et al., *Thermochimica Acta*, Vol. 297, 1997, pp. 85–92.
[8] Ghule, A., et al., *Thermochimica Acta*, Vol. 371, 2001, pp. 127–135.
[9] Chang, H., et al., *Materials Chemistry and Physics*, Vol. 58, 1999, pp. 12–19.
[10] Ghule, A., et al., *Inorganic Chemistry*, Vol. 40, 2001, pp. 5917–5923.
[11] Baskaran, N., et al., *Materials Chemistry and Physics*, Vol. 77, 2003, pp. 889–894.
[12] Murugan, R., et al., *Journal of Physics: Condensed Matter*, Vol. 12, 2000, pp. 677–700.
[13] Chang, H., et al., *Thermochimica Acta*, Vol. 374, 2001, pp. 45–49.
[14] Murugan, R., et al., *Thermochimica Acta*, Vol. 346, 2000, pp. 83–90.
[15] DuBois, C., et al., *30th NATAS Conference Proceedings*, 2002, pp. 214–219.

Journal of ASTM International, January 2005, Vol. 2, No. 1
Paper ID JAI12791
Available online at www.astm.org

Nancy L. Redman-Furey,[1] *Michael L. Dicks,*[2] *Jane Godlweski,*[2] *Dana C. Vaughn,*[3] *and Wendy J. Collins*[3]

The Role of TGA-DTA in the Initial Evaluation of the Solid State Forms for Pharmaceutical New Chemical Entities, Part 1: Evaluation of Pure Forms

ABSTRACT: TGA-DTA plays a central role in the strategy outlined for early evaluation of the solid state forms available to pharmaceutical new chemical entities. At this early stage of development, compound and time are typically at a premium, so a successful strategy requires making the best possible use of the materials and time available. In addition, because of time and compound limitations, the goal of a solid state investigation is focused upon early stage objectives rather than development of a complete understanding of all available solid state forms. The TGA-DTA is well suited to addressing these needs.

KEYWORDS: TGA, DTA, hydrate, thermal analysis, hygroscopicity, pharmaceutical

Introduction

Early in the drug development process, supply of the New Chemical Entity (NCE) and time to characterize the NCE are often limited. In this environment, successful strategies focus upon rapidly obtaining as much information as possible from limited samples. If the solid state form progressed from Research appears to meet the immediate Development goals, initial solid state studies focus upon: identification of the form(s) in use, characterization of the properties of the form(s) in use, identification of any special handling or drying requirements, and monitoring for lot to lot differences in solid state form.

The strategy developed in the authors' laboratory allows for lot-to-lot comparison and solid state identification using as little as 5–15 mg per lot, when the solid state forms present as pure phases. Sequentially evaluating samples via microscopy, then x-ray powder diffraction (XRPD), and lastly thermogravimetry combined with differential thermal analysis (TGA-DTA) ensures a visual record of a representative sample, conservation of sample by reuse of the XRPD sample, and placement of the destructive test last in the sequence. Initial assignment of the solid state form can often be made after these tests have been completed, assuming the sample has presented as a single phase. Confirmation is accomplished via a quick hygroscopicity challenge followed by TGA-DTA evaluation of the stressed samples.

Comprehensive reviews of the use of thermal analysis for solid state evaluation [1] and detailed discussions of pharmaceutical solid state issues [2] are already available to the reader. This paper describes a laboratory strategy for including TGA-DTA as a tool for solid state evaluation and provides examples for use of the data provided by the TGA-DTA in coordination with additional analytical techniques.

Manuscript received 24 May 2004; accepted for publication 15 September 2004; published January 2005.
[1] Principal Scientist, P&G Pharmaceuticals, Inc., Norwich, New York, 13815.
[2] Principal Researcher, P&G Pharmaceuticals, Inc., Norwich, New York, 13815.
[3] Senior Researcher, P&G Pharmaceuticals, Inc., Norwich, New York 13815.

Nitrofurantoin and PGE-551435 are presented as examples of the type of solid state variation observed during early development. Nitrofurantoin provides an example of a process that yields reproducibly a single, easy-to-handle phase while PGE-551435 presents an example of issues with both phase differences and sample drying. The samples used in the study are meant to represent the type of variation that may be observed during early development and to illustrate the utility of the proposed test strategy and are not meant to represent the first four development lots for these compounds.

Experimental

Microscopy

Sample slides were prepared using Cargille Meltmount 1.662 (Aroclor 5442 replacement) mounting medium. Photomicrographs were recorded from a Nikon e600 POL Scope and an Optronics 3-chip color camera using brightfield illumination.

X-ray Powder Diffraction

Powder diffraction patterns were collected using a Bruker D5000 x-ray diffractometer equipped with a position sensitive detector (PSD), from 2–40° 2θ, with detector slits set at ± 1 tic and a scan time of 0.2 s per step and a step size of 0.02° 2θ.

Thermogravimetry/Differential Thermal Analysis

Samples (5–10 mg) were run at 5°C /min from room temperature to onset of degradation under a dry nitrogen purge using a Seiko TG/DTA 220 SSC/5200. All of the samples were manually loaded rather than using an autosampler. Standard Seiko aluminum sample pans were used without covers. The TG/DTA was temperature calibrated using high purity (>99.999 %) gallium and NIST tin. The balance operation was confirmed using class I weights.

Hygroscopicity Challenge

Samples (in thin layers) were placed into a desiccator and into an 85 % RH chamber for 1–3 days, withdrawn, and evaluated by TGA-DTA. The 85 % RH chamber was prepared using a saturated KCl solution.

Samples

The samples used in this study are representative of the types of samples observed during the production of early development lots for Nitrofurantoin and PGE-551435. The four Nitrofurantoin samples were taken from P&GP commercial scale production lots because authentic samples from early development were not available for use in this study. The PGE-551435 samples were taken from both early development lots and from solvation screening studies to provide the greatest variety of examples with the fewest number of samples.

Results and Discussion

Nitrofurantoin

Nitrofurantoin (1-[[(5-nitro-2-furanyl)methylene]amino]-2,4-imidazolidinedione) is an antibacterial agent, developed in the 1950s and still prescribed today, for urinary tract infections under the trade names of Macrodantin, Macrobid, and Furadantin. The ability of Nitrofurantoin to form polymorphs and solvates has been discussed frequently in the literature [3–6].

In this study, four lots of Nitrofurantoin were examined, and each exhibited similar habits, x-ray powder diffraction patterns and TGA-DTA thermal curves. As seen in Fig. 1, thin, smooth acicular particles were observed for all samples. Matching crystal habit, while often indicative of a consistent solid state form, is neither a requirement nor a guarantee of identical solid state form.

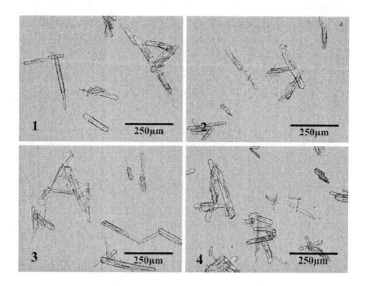

FIG. 1—*Matching photomicrographs for four lots of Nitrofurantoin.*

The powder diffraction data provided confirmation that all lots shared a single, crystalline solid state form. As shown in Fig. 2, the diffraction patterns for all four lots matched each other. Due to preferred orientation caused by the acicular particles, relative intensities for the diffraction patterns varied, but the peak positions for all lots were identical matches.

The TGA-DTA data provided identification of the solid state form. As shown in Fig. 3, all four lots exhibited a single, rapid weight loss with an associated sharp dehydration endotherm. The magnitude of the weight loss varied from 7.0–7.1 % across the lots, in excellent agreement with theory for monohydrate (7.03 %). In each case, the dehydration started above 100°C and was complete by 121–123°C. Dehydration at this temperature and with the rapidity shown in both the TGA and DTA curves is consistent with dehydration from a lattice type hydrate [7,8].

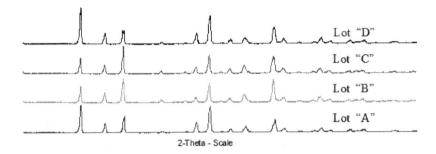

FIG. 2—*Matching x-ray powder diffraction patterns for four lots of Nitrofurantoin.*

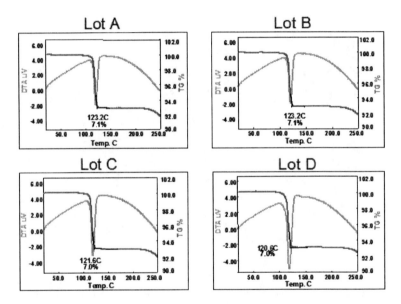

FIG. 3—*Matching TGA-DTA curves for four lots of Nitrofurantoin.   Thermal curve is consistent with loss of one mole of lattice type water of hydration.*

The hygroscopicity challenge confirmed the initial assignment of lattice monohydrate to the Nitrofurantoin samples. As shown in Fig. 4, exposure to either desiccation or 85 % RH failed to change the solid state form or the hydration level of the sample as indicated by unchanged TGA-DTA profiles. Under similar conditions, a channel type hydrate would be expected to lose water of hydration under desiccation and possibly gain under conditions of high humidity.

As Received

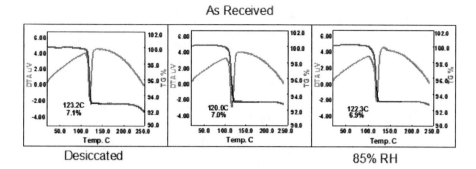

Desiccated   85% RH

FIG. 4—*Comparison of Nitrofurantoin as received and after desiccation and exposure to 85 % RH. No change is observed in the thermal curves.*

Nitrofurantoin Conclusions

Sequential use of light microscopy, x-ray powder diffraction, and TGA-DTA provided rapid confirmation that all four samples represented a single solid state form. It was the TGA-DTA data that identified the form as a lattice type monohydrate. This assignment was confirmed by the hygroscopicity challenge followed by additional TGA-DTA evaluation of the humidity and desiccation stressed samples. As expected for a lattice hydrate, no change in water content or solid state form was observed with changes in relative humidity.

This example demonstrated the utility of the combined techniques to quickly survey compounds produced during early development, but it also illustrates that this approach is not a substitute for full solvation state and polymorphism screens. In this case, the isolation process produced a single monohydrate form. Additional forms not observed in this particular process are available for Nitrofurantoin, with at least two monohydrates and two anhydrates cited in the literature [3–6].

PGE-551435

PGE-551435 is a development candidate under consideration as an antibacterial agent. The compound was initially isolated and evaluated as the free base. Descriptions of four different hydrate states for the free base have been described previously [9].

The four samples studied exhibited two distinct habits (Fig. 5). Samples 1 and 2 contained thick tablets of varying shape and irregularly-shaped flakes. Samples 3 and 4 contained crystals that were predominantly thin and four-sided, often loosely aggregated.

Similarly, two different XRPD patterns were observed for the samples (Fig. 6). Lots 1 and 2 exhibited matching patterns that were different from the matching pattern observed for Lots 3 and 4.

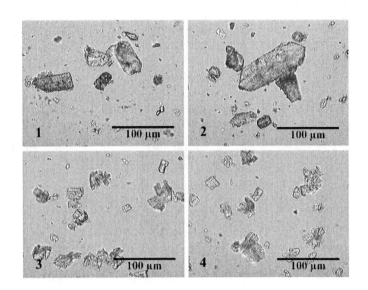

FIG. 5—*Photomicrographs illustrate differences between lots of PGE-551435.*

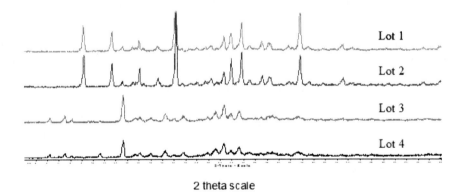

2 theta scale

FIG. 6—*Powder diffraction patterns for PGE-551435 indicate a similarity between Lots 1 and 2 and between Lots 3 and 4.*

The TGA-DTA profiles (Fig. 7) could be grouped into two types with total weight loss differences within each group. Lots 1 and 2 comprised one group and Lot 3 and 4 the other. In the first group, small weight losses (0.5 % and 1.5 %) were observed with no thermal events observed in the DTA trace prior to melt and degradation above 200°C. Such small, gradual weight losses are consistent with the presence of residual solvents, but the 1.5 % weight loss (close to theory for a ¼ hydrate) could have also indicated the presence of a channel type hydrate. Lots 3 and 4 exhibited higher weight losses, 16.6 % and 5.9 %, respectively. The weight losses were complete prior to 100°C and corresponded approximately to theory for 1 mole (4.8 %) and 4 moles (16.8 %) of water. In addition to the melt and degradation endotherm above 200°C, both lots exhibited an exotherm near 125°C. This exotherm is consistent with recrystallization following dehydration.

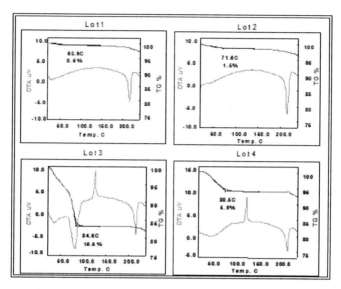

FIG. 7—*TGA-DTA curves for PGE-551435 illustrate differences in volatiles loss between lots.*

The combined evidence of microscopy, XRPD, and TGA-DTA separated the four samples into two groups. Additional information provided by the TGA-DTA suggested that Lots 1 and 2 were either anhydrates or channel type hydrates with very low hydration levels. Lots 3 and 4 appeared to be channel type hydrates that exhibited variable water content, samples containing variable amounts of residual solvent, or both. In addition, the recrystallization exotherm exhibited by Lots 3 and 4 suggested the presence of a stable anhydrate, providing further evidence that Lots 1 and 2 may be anhydrates. In a study separate from this initial survey, the material produced following the recrystallization exotherm was in fact demonstrated to match the phase present in Lots 1 and 2 [9].

The hygroscopicity challenge provided additional insight into the nature of these samples and the relationship between forms. Upon desiccation, Lot 1 remained unchanged, and the TGA response for Lot 2 decreased from 1.5 % to a 0.9 % loss (Fig. 8). Removal of most of the water from a channel hydrate often results in changes in the crystal lattice that can be observed microscopically or by XRPD. No such changes were observed by either technique. Weight loss over a broad temperature range and lack of evidence of change by microscopy or XRPD upon desiccation indicated that the TGA loss was most likely due to residual solvent from the crystallization process (toluene). Both lots gained approximately 10 % water upon exposure to 85 % RH. The TGA-DTA curves showed immediate, rapid loss of water upon start of heating, consistent with loss of surface water. A second endotherm in the DTA traces and a second weight loss step in the TGA curves (60–80°C) indicated loss of water from a different environment, very similar in appearance to the dehydration events observed for Lots 3 and 4. Powder diffraction data for samples taken from the humidity condition showed that while the initial phase of Lots 1 and 2 was still largely intact, the phase present in Lots 3 and 4 was also beginning to grow in. Taken collectively, the data indicated that Lots 1 and 2 were anhydrates containing varying amounts of residual solvent and that upon exposure to 85 % RH, the anhydrate had begun to convert to the hydrates phase observed in Lots 3 and 4.

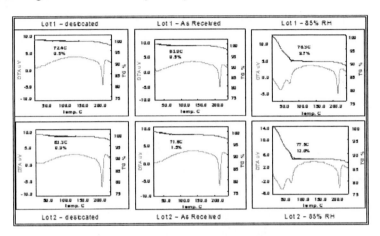

FIG. 8—*TGA-DTA curves for PGE-551435 Lots 1 and 2 hygroscopicity samples. Water gain and initial conversion to a hydrated form can be seen from the 85 % RH sample curves.*

The water content of Lots 3 and 4 dropped to 3.2 % and 2.9 %, respectively, as a result of desiccation (Fig. 9). This decrease in water content resulted in Lot 4 becoming hygroscopic enough to pick up water under ambient laboratory humidity conditions. This led to the unusual observation of an initial TGA weight gain for Lot 4 after desiccation. Loss of water under desiccation is consistent with channel type water of hydration. As reported in an earlier publication, desiccation to this moisture level resulted in an apparent decrease in crystallinity (as measured by XRPD) [9]. This is believed to be have been caused by a decrease in crystallite size caused by lattice instability and fracturing of the crystals as the water content dropped. This observation is also consistent with channel type water. Upon exposure to 85 % RH, Lots 3 and 4 gained water, exhibiting TGA weight losses of 23.4 % and 23.1 %, respectively. Even with this large an increase in water content, the XRPD data indicated no change in crystal lattice and no evidence of conversion to new phase(s). The ability of these two lots to desiccate to approximately the same low water value and to increase to approximately the same high water content upon exposure to 85 % RH indicated that the difference in water content between the two as received was likely the result of different drying histories and that, as the microscopy and XRPD data indicated, Lots 3 and 4 represented the same solid state form.

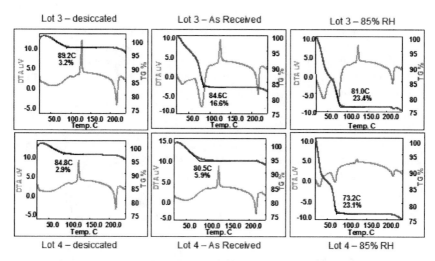

FIG. 9—*TGA-DTA curves for PGE-551435 Lots 3 and 4 hygroscopicity samples. Changes in water content over the RH conditions are consistent with a channel type hydrate.*

PGE-551435 Conclusions

Sequential use of light microscopy, x-ray powder diffraction, and TGA-DTA provided rapid confirmation that the four samples represented two different solid state forms. TGA-DTA data and the hygroscopicity challenge identified the form of Lots 1 and 2 as an anhydrate. The TGA-DTA data identified the form of Lots 3 and 4 as a channel type hydrate. Using XRPD in addition to TGA-DTA to evaluate samples generated in the hygroscopicity challenge, the anhydrate was observed to convert to the channel hydrate under high humidity conditions. Additionally, the channel hydrate was observed to spontaneously convert to the anhydrate upon heating. TGA-DTA data for the channel hydrate illustrated the wide range of water contents (3–23 %) available to the channel hydrate, highlighting the need for caution in handling, storing, and drying this form.

This example demonstrates the utility of these combined techniques to quickly survey compounds produced during early development and to provide identification of hydration type and level. Although the test strategy is designed as a quick survey rather than a full solid state screen, it can, as demonstrated here, provide information regarding the interrelationship and relativity stability of the different forms examined. In addition, the hygroscopicity challenge, in combination with the TGA-DTA, confirmed the identity of solid state forms and, in the case of the channel hydrate, provided information about the range of water contents available to that particular form.

Conclusions

TGA-DTA plays a central role in the initial evaluation of the solid state form for pharmaceutical new chemical entities. In the overall testing strategy described here, TGA-DTA was the technique responsible for identification of hydrate level (stoichiometry) and type (channel versus lattice). In some cases the TGA-DTA curves also provided early evidence for phase conversions. The TGA-DTA, in combination with hygroscopicity studies, enabled rapid screens of hydrate stability to differing ambient humidities. Collectively this data facilitated an early understanding of the nature of the solid state forms of the samples and identification of needs for special handling, storage, and/or processing conditions.

References

[1] Giron, D., "Thermal Analysis and Calorimetric Methods in the Characterisation of Polymorphs and Solvates," *Thermochimica Acta*, Vol. 248, 1995, pp. 1–59.

[2] *Polymorphism in Pharmaceutical Solids*, Marcel Dekker and H. G. Brittain, Eds., New York, 1999.

[3] Garti, N. and Tibika, F., "Habit Modifications of Nitrofurantoin Crystallized from Formic Acid Mixtures," *Drug Development and Industrial Pharmacy*, Vol. 6, No. 4, 1980, pp. 379–398.

[4] Marshall, P. V. and York, P., "Crystallisation Solvent Induced Solid-State and Particulate Modifications of Nitrofurantoin," *International Journal of Pharmaceutics*, Vol. 55, 1989, pp. 257–263.

[5] Pienaar, E. W., Caira, M. R., and Lötter, A. P., "Polymorphs of Nitrofurantoin. I. Preparation and X-Ray Crystal Structures of Two Monohydrated Forms of Nitrofurantoin," *Journal of Crystallographic and Spectroscopic Research*, Vol. 23, No. 9, 1993, pp. 739–744.

[6] Caira, M. R., Pienaar, E. W., and Lötter, A. P., "Polymorphism and Pseudopolymorphism of the Antibacterial Nitrofurantoin," *Mol. Cryst. Liq. Cryst.*, Vol. 279, 1996, pp. 241–264.

[7] Morris, K. R., "Structural Aspects of Hydrates and Solvates," *Polymorphism in Pharmaceutical Solids*, Marcel Dekker and H. G. Brittain, Eds., New York, 1999, pp. 125–182.

[8] Redman-Furey, N. and Collins, W., "Thermoanalytical Characterization of Pharmaceutical Materials: Small Molecule Applications," *American Pharmaceutical Review*, Vol. 5 No. 1, 2002, pp. 102–107.

[9] Dicks, M. L., Parsons, D. M., and Redman-Furey, N. L., "Concurrent Use of Thermal Analysis and X-Ray Powder Diffraction to Characterize Hydration States," *Proceedings of the 30th North American Thermal Analysis Society Conference*, B&K Publishing, K. J. Kociba, Ed., 2002, pp. 197–202.

Journal of ASTM International, January 2005, Vol. 2, No. 1
Paper ID JAI12792
Available online at www.astm.org

Nancy L. Redman-Furey,[1] Michael L. Dicks,[2] Jane Godlweski,[2] Dana C. Vaughn,[3] and Wendy J. Collins[3]

The Role of TGA-DTA in the Initial Evaluation of the Solid State Forms for Pharmaceutical New Chemical Entities, Part 2: Evaluation of Mixed Forms

ABSTRACT: TGA-DTA plays a central role in the strategy outlined for early evaluation of the solid state forms available to pharmaceutical new chemical entities. Understanding of the solid state forms becomes more difficult when individual samples present as mixed forms, especially when it is not immediately recognized that the samples represent a mixture. In this study, TGA-DTA, in combination with light microscopy and powder X-ray diffraction, provided immediate evidence that samples represented mixed solid state forms. The initial assessment was made using as little as 5 mg of sample. Hygroscopicity challenges provided further proof for mixed forms. To make a definite assignment of the solid state forms present, isolation of pure phases of the suspected individual forms was necessary. Success of this testing strategy is illustrated using an example of mixed salt stoichiometry and mixed hydration states. A hierarchy is suggested for efficient isolation efforts when a complex mixture of solid state samples is present.

KEYWORDS: TGA, DTA, hydrate, thermal analysis, hygroscopicity, pharmaceutical, Risedronate

Introduction

Early in the drug development process, supply of the New Chemical Entity (NCE) and time to characterize the NCE are often limited. In this environment, successful strategies focus upon rapidly obtaining as much information as possible from limited samples. Initial solid state studies often focus upon: identification of the form(s) in use, characterization of the properties of the form(s) in use, identification of any special handling or drying requirements, and monitoring for lot-to-lot differences in solid state form. Efforts to identify and characterize the solid state form become complicated when early drug lots present as mixed forms. Comprehensive reviews are available that discuss the need for solid state evaluation of pharmaceutical materials and describe the use of thermal analysis in the characterization of solvates and polymorphs [1,2].

The strategy developed in the authors' laboratory allows for lot-to-lot comparison using as little as 5 mg per lot, for initial assessment of phase purity and identity. Sequentially evaluating samples via microscopy, then x-ray powder diffraction (XRPD), and lastly thermogravimetry, combined with differential thermal analysis (TGA-DTA), ensures a visual record of a representative sample and conservation of sample by reuse of the XRPD sample, and it places the destructive test last in the sequence. When initial test results suggest the presence of mixed phases within a single sample, preparation of pure phases is required to confirm phase identity for each component. A logical sequence for preparation and evaluation of pure phases is suggested and illustrated.

Manuscript received 24 May 2004; accepted for publication 15 September 2004; published January 2005.
[1] Principal Scientist, P&G Pharmaceuticals, Inc., Norwich, New York, 13815.
[2] Principal Researcher, P&G Pharmaceuticals, Inc., Norwich, New York, 13815.
[3] Senior Researcher, P&G Pharmaceuticals, Inc., Norwich, New York 13815.

Risedronate provides an example of the type of solid state variation observed during early development and demonstrates the need to be able to identify phases within mixed phase samples. The actual samples used in the study are meant to represent the type of variation that may be observed during early development and are not meant to represent the first four development lots of Risedronate.

Experimental

Microscopy

Sample slides were prepared using Cargille Meltmount 1.662 (Aroclor 5442 replacement) mounting medium. Photomicrographs were recorded from a Nikon e600 POL Scope and an Optronics 3-chip color camera using brightfield illumination.

X-ray Powder Diffraction

Powder diffraction patterns were collected using a Bruker D5000 x-ray diffractometer equipped with a position sensitive detector (PSD), from 2–40° 2θ, with detector slits set at ± 1 tic and a scan time of 0.2 s per step and a step size of 0.02° 2θ.

Thermogravimetry/Differential Thermal Analysis

Samples (5–10 mg) were run at 5°C/min from room temperature to onset of degradation under a dry nitrogen purge using a Seiko TG/DTA 220 SSC/5200. All of the samples were manually loaded rather than using an autosampler. Standard Seiko aluminum sample pans were used without covers. The TG/DTA was temperature calibrated using high purity (>99.999 %) gallium and NIST tin. The balance operation was confirmed using class I weights.

Hygroscopicity Challenge

Samples (in thin layers) were placed into a desiccator and into an 85 % RH chamber for 1–3 days, withdrawn, and evaluated by TGA-DTA and x-ray. The 85 % RH chamber was prepared using a saturated KCl solution.

Samples

The samples used in this sample are representative of the types of samples observed during the production of early development lots for Risedronate. The samples were obtained from very early development lots and from recrystallization studies using water and isopropanol.

Results and Discussion

Risedronate, [1-Hydroxy-2-(3-pyridinyl)ethylidene]bis[phosphonic] acid is the active in the osteoporosis drug Actonel. The thermal characterization of anhydrate, monohydrate, and hemi-pentahydrate forms of the monosodium salt of Risedronate has been described previously [3]. In addition to differing hydration states, the possibility of varying salt stoichiometry exists due to the multiple ionization sites available to bisphosphonates.

In this study, four lots of Risedronate were examined and found to exhibit different crystal habits, x-ray powder diffraction patterns, and TGA-DTA thermal curves from each other. As seen in Fig. 1, photomicrographs of the four lots show variation of particle habit within lot and lot-to-lot. Lot 1 contained sharp-edged monoclinic plates and elongated tablets with tapered edges in addition to irregular fragments. Sharp-edged acicular particles along with elongated tablets were observed in Lot 2. Lot 3 contained sharp-edged elongated flakes and irregular spheroidal particles. The elongated tablets were again observed in Lot 4 in addition to irregular flakes. Irregularly-shaped fragments were observed in all of the lots. The appearance of very different habits within an individual lot is consistent with: the isolation of differing chemical entities (the active and impurity), differing solid state forms of the same parent compound, or the same solid state form under differing nucleation conditions (such as degree of supersaturation) [4]. In this case, because of efforts to control and monitor the chemical purity, the first possibility would appear unlikely due to the relative amounts of the differing habits.

The powder diffraction data (Fig. 2) also indicated lot-to-lot differences. Some elements of the diffraction patterns were common to all the lots, while others appeared to be unique to a given lot. This result is consistent with the presence of mixed phases with a common phase between the samples.

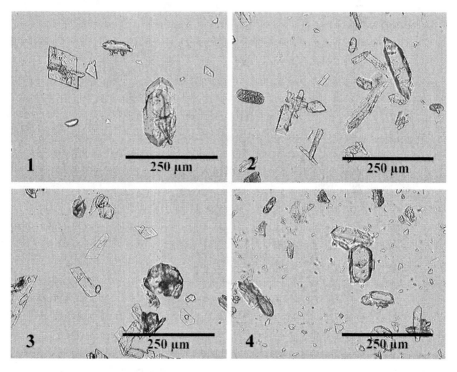

FIG. 1—*Photomicrographs exhibiting differing Risedronate crystal habits within and between lots.*

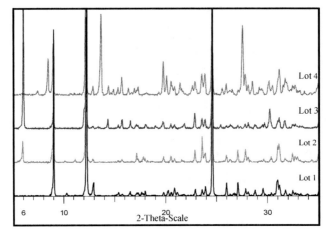

FIG. 2—*Differing powder diffraction patterns for each lot of Risedronate.*

The TGA-DTA data, in agreement with the microscopy and powder diffraction data, indicated the presence of mixed solid state forms (Fig. 3). Each of the profiles exhibited a weight loss step below 100°C as well as weight loss above 100°C. The weight loss step associated with dehydration below 100°C is consistent with loss of channel type water of hydration. Channel water is recognized to exist within channels or tunnels within the crystal lattice and may move with relative ease into or out of the crystal lattice, depending upon ambient conditions. Lattice water is understood to be isolated within the crystal lattice in such a manner that it cannot be removed without destroying the crystal lattice and as a result typically requires a temperature above the boiling point of water for removal as well as weight loss step(s) above 100°C [5,6]. Consequently, the presence of dehydration steps above and below 100°C indicated the existence of both channel type and lattice type waters of hydration. The higher temperature weight losses could have been ascribed to degradative weight loss without additional evidence to assign these transitions to dehydration. Previous experience with the class of compound (bisphosphonates) suggested that these losses could be dehydration, and this was confirmed by separate Karl Fischer experiments. Lots 1 and 3 exhibited single step weight losses of approximately 3 % up to 100°C, consistent with loss of less than one mole of channel type water of hydration. Both of these lots exhibited two step losses at temperatures above 150°C, consistent with loss of lattice or ion associated type water of hydration. The high temperature losses for Lot 3 were more clearly separated than those of Lot 1. Lot 2 exhibited approximately 8 % weight loss to 120°C, consistent with loss of at least one mole of channel water. Like Lots 1 and 3, an additional weight loss was observed above 150°C, consistent with loss of lattice water of hydration. Lot 4 initially exhibited a weight gain on the TGA curve, consistent with water gain due to the extreme hygroscopicity of the sample. (Although very dry nitrogen was used to purge the TGA-DTA, ambient lab humidity provided a water source for the sample as a result of opening the instrument to load the sample. The furnace volume is large enough that the ambient humidity was not immediately displaced by the nitrogen purge.) As the temperature of the scan increased, weight loss was observed for a 2 % loss prior to 100°C, consistent with dehydration of less than one mole of channel water. Like the other lots, an additional weight loss was observed above 150°C, consistent with dehydration of a lattice hydrate. The TGA-DTA evidence clearly indicated presence of both channel and lattice type waters of hydration within each lot. While the TGA-DTA curves contained some similar elements, the individual profiles obtained for each lot were unique. Lots 1, 3, and 4 exhibited approximately 10 % weight loss, consistent with theory for a monosodium dihydrate (10.6 %). Based upon TGA-DTA, XRPD, and microscopy data, a common component may have been present between lots, but the lots were clearly not identical. The 13.1 % weight loss observed for Lot 2 is consistent with theory for monosodium hemi-pentahydrate (12.9 %), but again, taken in total, the data suggested that this sample contained mixed forms.

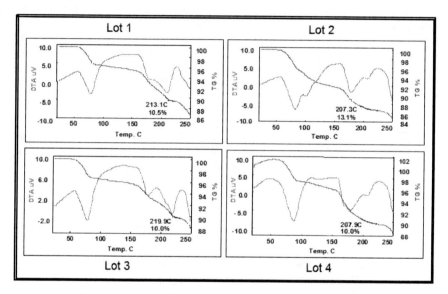

FIG. 3—*Differing TGA-DTA thermal curves for each lot of Risedronate. The pattern of weight losses suggests the presence of both lattice and channel type waters of hydration.*

The hygroscopicity challenge provided additional information related to hydration behavior. TGA-DTA curves of the stressed samples indicated loss or gain of water only in the temperature region corresponding to channel type water (< 100°C). No changes (such as complete loss of a given thermal event or shifts in temperature of thermal events) were observed that would indicate inter-conversion between forms. Lots 1 and 3 lost 3.2 % and 2.3 % water, respectively, under desiccation but did not change upon exposure to 85 % RH, remaining close to theory for the dihydrate. If it were not for the data suggesting that these lots are mixtures, a tentative assignment for dihydrate might have been made. Lot 2 lost 3.1 % water under desiccation and gained 1.0 % water at 85 % RH, changing to a hydration level consistent with a trihydrate. Lot 4 lost 2.3 % water under desiccation and gained 2.9 % upon exposure to 85 % RH. This water gain resulted in a change of moisture content corresponding to a change from dihydrate and hemi-pentahydrate. Loss of water in all samples was consistent with the assumption that at least part of the water of hydration in each sample existed in channel form. While each of the lots appeared to display one or more stoichiometric hydration levels, each lot behaved differently in the hygroscopicity challenge. This provided further evidence of the uniqueness of each sample but did not provide information to identify the solid state forms present.

With overwhelming evidence indicating that most, if not all, of the lots contained mixed forms, efforts turned toward isolation of pure forms for comparison. The hierarchy used in the authors' laboratory called for isolation of differing salt stoichiometries (if available) first, followed by isolation of differing solvation forms for each salt, and finally isolation of individual polymorphs for each solvation state of interest. While the monosodium salt of Risedronate was the targeted salt form, the multiple ionization sites available to the compound (pKas of: 1.6, 2.2, 5.9, 7.1, 11.7) suggested that the free acid and disodium salt could be isolated if the pH was not controlled adequately. Sodium assays for the samples indicated a low result for Lot 1, high assay for Lot 2, and a match to theory for monosodium for Lot 3 and 4. Crystallization experiments conducted at carefully controlled pH conditions, but otherwise matching the process chemistry conditions, yielded samples that when assayed for sodium matched the expected values for disodium salt, monosodium salt, and free acid. These pure salt forms were then used for comparison to the four initial lots.

Taking into consideration the low sodium assay obtained for Lot 1 and the high sodium assay obtained for Lot 2, both lots were compared to the TGA-DTA responses obtained for free acid, monosodium salt, and disodium salt. As shown in Fig. 4, the TGA-DTA profile for Lot 1 represented a composite of the thermal responses obtained for free acid and monosodium salt, consistent with a sodium assay lower than expected for the monosodium salt. Similarly, the TGA-DTA curve Lot 2 (Fig. 5) represented a composite of monosodium and disodium salts as might have been predicted based upon the high sodium assay. These results were also confirmed by XRPD.

As shown in Fig. 6, neither Lot 3 nor 4 matched the TGA-DTA curves obtained for the monosodium salt isolated from the controlled pH study, although the sodium assay for both lots was consistent with that expected for a monosodium salt. This result necessitated an evaluation of the hydration states available to the monosodium salt, the second level in the hierarchy of solid state evaluation. Isolation efforts took into consideration the process histories for Lots 3 and 4, anticipated temperature, and solvent extremes for the crystallization step, varying degrees of supersaturation and rate of isolation. Upon completion of these experiments, a monosodium monohydrate (5.6 % water) and monosodium hemi-pentahydrate (12.9 % water) were isolated as pure forms and demonstrated to be the only hydrates expected under normal process conditions. The hemi-pentahydrate, as the form in equilibrium with water at room temperature, was the form of the monosodium salt initially isolated. As shown in Fig. 7, the thermal curves obtained for Lot 3 can be understood as the composite of the curves expected for the monohydrate and hemi-pentahydrate forms of the monosodium salt.

For Lot 4 (Fig. 8) the comparison was not so obvious. While similar in appearance to the hemi-pentahydrate, the water loss of 10.0 % more closely matched the theory for dihydrate than hemi-pentahydrate. The hygroscopicity challenge for Lot 4 provided the answer (Fig. 9). The low water value for Lot 4 as received was due to over-drying of the channel portion of the hydrate. After exposure to high humidity, the total water loss, TG/DTA profile, and diffraction pattern for Lot 4 all matched that of the hemi-pentahydrate.

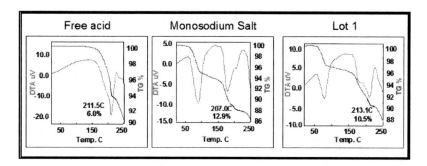

FIG. 4—*The thermal curve obtained for Lot 1 appears to be a composite of the thermal curves obtained for free acid and monosodium salt.*

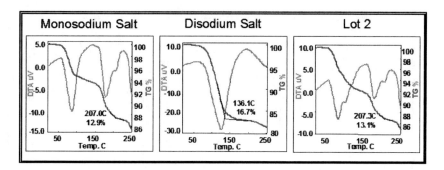

FIG. 5—*The thermal curve obtained for Lot 2 appears to be a composite of the thermal curves obtained for monosodium salt and disodium salt.*

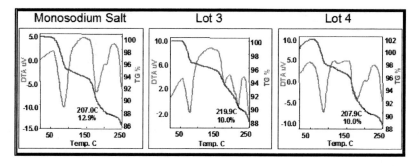

FIG. 6—*The thermal curves obtained for Lots 3 and 4 do not match that obtained for monosodium salt.*

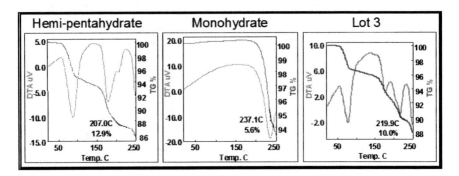

FIG. 7—*The thermal curve obtained for Lot 3 appears to be a composite of the thermal curves obtained for monosodium salt hemi-pentahydrate and monohydrate.*

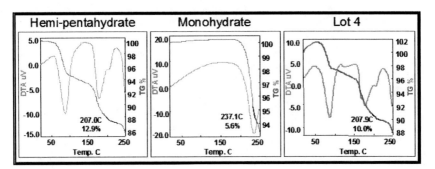

FIG. 8—*The thermal curve for Lot 4 does not appear to be a composite of the curves for hemi-pentahydrate and monohydrate.*

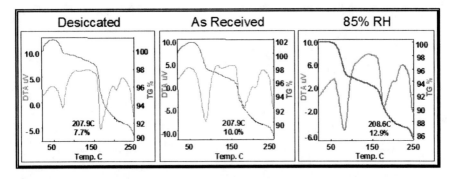

FIG. 9—*The hygroscopicity results for Lot 4 indicate a match for hemi-pentahydrate upon full hydration of the match. The sample as received was over-dried.*

Conclusion

The presence of mixed salt and/or solvate forms in early development samples complicates solid state identification and can slow development if not discovered. Use of TGA-DTA with the assistance of microscopy and powder x-ray diffraction enabled early detection of mixed forms. In the examples provided here, use of Karl Fischer or loss on drying (LOD) alone to determine water may have led to the erroneous assignment of dihydrate to three of the samples examined as a result of a coincidental matching of water level to a stoichiometric amount. In addition, use of Karl Fischer to monitor water would not have enabled early identification of the two types of water of hydration (channel and lattice) present in these samples as accomplished by TGA-DTA. As shown in these samples, use of LOD may not have been practical or possible due to the very high temperature of the final dehydration and lack of separation of dehydration from onset of degradation.

Identification of the individual components depended upon isolation of pure phases. The isolation strategy illustrated in this study was a tiered approach that focused first upon isolation of individual salts of defined stoichiometry, followed by isolation of individual hydrates for a given salt. The final tier, had it been necessary, would have been isolation of individual polymorphs for given hydrate. TGA-DTA played a central role in the evaluation of the solid state form both before and after pure phases had been isolated.

The TGA-DTA provided information relating to both hydration level and nature of the hydrate that was not available from the other techniques. Early identification of the existence of channel type water is necessary for adequate control of handling and drying for the material. The content mobile water present in channel hydrates can readily change, as was the case for Lot 4. If the relative ease for water content change is not understood, changes in the compound water content could inadvertently occur due to inadequate controls in storage and/or handling.

References

[1] *Polymorphism in Pharmaceutical Solids*, Marcel Dekker and H. G. Brittain, Eds., New York, 1999.

[2] Giron, D., "Thermal Analysis and Calorimetric Methods in the Characterisation of Polymorphs and Solvates," *Thermochimica Acta*, Vol. 248, 1995, pp. 1–59.

[3] Redman-Furey, N. L., Collins, W. J., and Burgin, M. A., "Thermoanalytical Characterization of the Hydration States of Risedronate," *Proceedings of the 30th North American Thermal Analysis Society Conference,* K. J. Kociba, Ed., B&K Publishing, 2002, pp. 733–738.

[4] Davey, R. and Garside, J., "Crystal Morphology," *From Molecules to Crystallizers, An Introduction to Crystallization,* Oxford University Press, Oxford, 2000, pp. 36–43.

[5] Morris, K. R., "Structural Aspects of Hydrates and Solvates," *Polymorphism in Pharmaceutical Solids*, Marcel Dekker and H. G. Brittain, Eds., New York, 1999, pp. 125–182.

[6] Redman-Furey, N. L. and Collins, W. J., "Thermoanalytical Characterization of Pharmaceutical Materials: Small Molecule Applications," *American Pharmaceutical Review*, Vol. 5, No. 1, 2002, pp. 102–107.

Journal of ASTM International, July/August 2005, Vol. 2, No. 7
Paper ID JAI12790
Available online at www.astm.org

Adrienne S. Bigalow Kern,[1] *Wendy J. Collins,*[1] *R. Thomas Cambron, Ph.D.,*[2] *and Nancy L. Redman-Furey, Ph.D.*[3]

Use of a TG/DTA/Raman System to Monitor Dehydration and Phase Conversions

ABSTRACT: Use of a TG/DTA/Raman system enabled real time collection of Raman spectra during the dehydration of Risedronate, the active pharmaceutical ingredient in the osteoporosis drug Actonel. Raman spectra collected during the dehydration of this material show a crystal lattice adjustment below the boiling point of water facilitating interpretation of the TG/DTA thermal profile. Raman spectra also revealed recrystallization processes subsequent to loss of lattice water that were not evident in the thermal analysis profile. Spectral evidence showing the formation and disappearance of a second hydrate form are observed during the dehydration of this material. In addition, spectral evidence indicating the presence of an anhydrate form of this material are observed prior to thermal induced degradation of this material.

KEYWORDS: Thermal Analysis, TGA, TG/DTA, Raman, Raman Spectroscopy, Hydration state

Introduction

Thermogravimetric Analysis (TGA) and Differential Thermal Analysis (DTA) are well-established techniques in the pharmaceutical industry used to characterize the hydration state of an active pharmaceutical ingredient (API). The combination of these two thermal analysis techniques enables quantitative determination of hydration state and provides information related to the type of water (channel or lattice) present in the API. Distinct thermal events observed in the DTA thermal profile provide valuable information for the characterization of hydrated materials. However, detecting distinct thermal events from a DTA thermal profile can be challenging. In addition, the origin of temperature-induced mass loss is often difficult to identify from a thermal mass loss profile alone. Therefore, additional analytical measurements have been coupled to TGA instrumentation to facilitate interpreting thermal mass loss profiles. Early examples include coupling FTIR and MS to TGA for monitoring gasses evolved during the experiment [1–3]. Characterization of gasses evolved during the thermal experiment enable identification of degradation products [2,3]. In addition to monitoring evolved gasses, a variety of analytical measurements has been used to characterize mass loss in these thermal experiments off-line [3,4]. In these experiments, the thermal experiment is disrupted and the sample removed for analysis in a laboratory. Although this approach has proven valuable for direct analysis of the sample during a thermal experiment, changes in sample composition during preparation for laboratory analysis limit the practicability of this approach.

Manuscript received 24 May 2004; accepted for publication 8 December 2004; published July 2005. Presented at ASTM Symposium on Elemental Analysis of Fuels and Lubricants: Recent Advances and Future Prospects on 6-8 December 2004 in Tampa, FL.
[1] Senior Researcher, Procter & Gamble Pharmaceuticals, P.O. Box 191, State Route 320, Norwich, NY 13815.
[2] Principal Scientist, Procter & Gamble Pharmaceuticals, P.O. Box 191, State Route 320, Norwich, NY 13815.
[3] Senior Scientist, Procter & Gamble Pharmaceuticals, P.O. Box 191, State Route 320, Norwich, NY 13815.

Raman spectroscopy has been exploited for monitoring temperature-induced changes in composition and phase transformations [5–16]. Early applications involved continuous monitoring of the sample in a temperature controlled oven. This technique was referred to as thermo-Raman spectroscopy (TRS). Subsequent reports have demonstrated feasibility for conducting TRS measurements in commercially available TGA instrumentation [17–19]. Direct coupling of Raman to a commercial TGA instrument provides several advantages compared to TRS conducted with ovens. One key advantage is consistent sampling conditions for individual measurements without disruption of the thermal experiment. Recent reports have demonstrated applications of TRS for monitoring thermal decomposition and temperature induced dehydration and phase transformations [20–22]. Experiments summarized in this report utilize this approach for monitoring a TG/DTA. Feasibility for simultaneous collection of DTA thermal profiles utilizing commercial instrumentation and TRS has been demonstrated.

The dehydration of Risedronate, [1-Hydroxy-2-(3-pyridinyl)ethylidene]bis[phosphonic] acid, the active ingredient in the osteoporosis drug Actonel, was characterized by coupling a Raman spectrometer directly to a commercial TG/DTA. Three different solid-state forms of Risedronate are used to demonstrate the informing power of this hyphenated analytical methodology [23]. This report demonstrates how coupling Raman spectroscopy to an established thermal technique enhances interpretation of thermal events.

Experimental

Thermogravimetric Analysis

A commercially available Seiko TG/DTA 220 was modified to allow the Raman probe unobstructed access to the sample. The modification consisted of imbedding a quartz window within the ceramic core of an existing furnace. A secondary quartz sleeve was needed to maintain the same heating effect as a traditional furnace assembly. The operational range is reduced as a result of these modifications. The range of the traditional furnace is from room temperature to 1050°C. As a result of imbedding the quartz window, the custom furnace's upper range is reduced to 550°C. Most pharmaceutical API materials do not contain any water above approximately 250°C. Therefore, the reduced upper temperature limit of the modified furnace assembly does not impact the application of this technology for most pharmaceutical applications. A picture of a traditional and modified furnace assembly is shown in Fig. 1a.

Scans were made at 5°C/min using a dry nitrogen purge. The instrument was temperature calibrated using gallium and NIST traceable tin. Mass calibration was confirmed using Class I weights.

Raman Spectroscopy

Raman spectroscopy is an established technique for studying fundamental molecular vibrations [24,25]. The Raman effect is based on inelastic scattering of laser light from the oscillating electron cloud around a molecule. Similar to mid-infrared spectroscopy, the resulting spectrum probes fundamental molecular vibrations. In these experiments, Raman spectra were collected utilizing a Kaiser Raman Holoprobe spectrometer (785 nm laser excitation) coupled to a filtered fiber optic probe head. Laser illumination of the sample was achieved with a non-contact (2.5 in. focal length) optical element attached to the probe head. This configuration enabled collection of spectra through the imbedded quartz window. The Raman probe head was positioned using multiple translational stages enabling alignment of the laser beam with the sample. A picture of the coupled instrument is shown in Fig. 1*b*. Suitable Raman acquisition exposure intervals depend upon the Raman scattering cross-section of the sample. In these experiments, a 30 s Raman acquisition exposure and a 30 s cosmic ray filter produced spectra with suitable signal-to-noise for interpretation [26]. The TG/DTA and Raman were started simultaneously with the Raman collecting one spectrum for every 5° increase in temperature.

(a)   (b)

FIG. 1—(a) *Traditional TG/DTA furnace on the right, modified furnace with quartz window shown on* (b) *TG/DTA/Raman setup.*

The influence of localized sample heating due to intense laser radiation exposure was characterized prior to collecting dehydration data. A sample was placed in the TG/DTA and exposed to the focused Raman laser while collecting the corresponding TG/DTA signal at room temperature. A significant DTA deflection (~20µV) was observed at room temperature utilizing the maximum power output of the Raman laser. Therefore, the laser intensity was reduced by approximately forty percent to minimize the DTA deflection to <10µV. Sequential Raman spectra collected at this reduced laser intensity under isothermal conditions in the TG/DTA instrument reveal a small amount of sample heating due to exposure of the sample to the Raman laser. However, this laser induced heating has a negligible impact on Raman spectra collected in these experiments. In addition, comparing TG/DTA experiments conducted with and without the Raman laser energized indicate localized heating from the Raman laser has a negligible impact on the temperatures of the dehydration events.

Materials—Three different solid-state forms of Risedronate were used in these experiments. These materials represent different hydration states of Risedronate. They include the hemi-pentahydrate, monohydrate, and anhydrate forms of Risedronate. A pseudomorph of the hemi-pentahydrate was prepared by desiccation of hemi-pentahydrate material to remove channel water from the crystal lattice. A production lot of the hemi-pentahydrate form of Risedronate was obtained from P&G Pharmaceuticals, Inc. The normal production material was slurried in a water/isopropanol mixture at 65°C for several days to prepare the monohydrate phase. The overdried hemi-pentahydrate was prepared by placing hemi-pentahydrate in a dessicator for several days. To prepare the anhydrate, the water of hydration of the normal production material was azeotroped off in N, N-dimethylformamide. Additional studies have been conducted to characterize and support the identification of these different forms of Risedronate [27]. Results of these characterization studies are beyond the scope of this report.

Results and Discussion

The active pharmaceutical ingredient (API) in the osteoporosis drug Actonel is Risedronate [1-Hydroxy-2-(3-pyridinyl)ethylidene]bis[phosphonic] acid. Three different hydration states (hemi-pentahydrate, monohydrate, and anhydrate) and overdried hemi-pentahydrate have been characterized by a combination of spectroscopy and thermal analysis. The hemi-pentahydrate form of Risedronate is a mixed hydrate containing both channel water and lattice water. The channel water can be removed from the crystal at temperatures below the boiling point of water to produce a partially hydrated, or "overdried," pseudomorph of the hemi-pentahydrate. Destruction of the crystal lattice results from removal of the lattice water from the overdried hemi-pentahydrate at temperatures above the boiling point of water. The monohydrate form of Risedronate contains only lattice water, and the anhydrate form does not contain water. The crystal structure for each hydrate is unique for each form. Differences in water content and location for each hydrate produce unique Raman spectra for each form and for the pseudomorph. The primary source of spectral differences between the different forms is due to the impact of hydrogen bonding on the fundamental vibrational states for each hydrate. Spectral features unique to each form are observed in the CH stretching region (2800 cm^{-1} to 3200 cm^{-1}) and the corresponding CH deformation region below 1500 cm^{-1}. The spectral range between 700 cm^{-1} and 1000 cm^{-1} contains unique spectral features for form of Risedronate that correspond to aromatic CH deformations. All spectra illustrated in this report are displayed in this spectral range to facilitate visual interpretation. Representative Raman spectra of each forms and the psedudomorph of Risedronate are illustrated in Fig. 2.

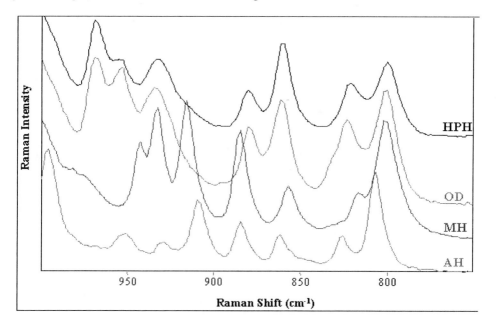

FIG. 2—*Representative Raman spectra of hemi-pentahydrate (HPH), overdried hemi-pentahydrate (OD), monohydrate (MH), and anhydrate (AH) forms of Risedronate.*

The TG/DTA thermal profile collected during the dehydration of hemi-pentahydrate is illustrated in Fig. 3.   The first dehydration event observed in Fig. 3 at approximately 75°C represents removal of 1.5 mole equivalent of channel water from the crystal lattice. This channel water is weakly associated with the crystal lattice and is removed from the crystal at a temperature below the boiling point of water.   Raman spectra collected during this initial dehydration event between 25°C and 135°C are illustrated in Fig. 4.   Removal of the channel water from the hemi-pentahydrate produces Raman spectra different from those obtained from the fully hydrated species. As a result, this dehydration process can be directly monitored by Raman spectroscopy.   At 65°C, the corresponding Raman spectrum matches the reference spectrum for overdried hemi-pentahydrate and does not display characteristics of either the fully hydrated hemi-pentahydrate or monohydrate forms of Risedronate.

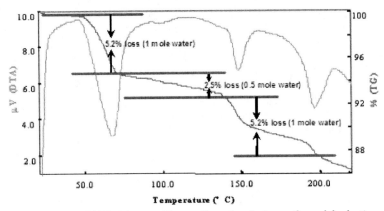

FIG.   3—*TG/DTA   (5°C/min)   profile   collected   during   the   dehydration   of   the hemipentahydrate form of Risedronate. (Exotherm up.)*

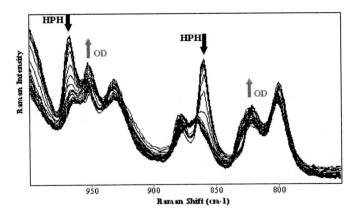

FIG. 4—*Raman   spectra   collected   between   25°C   and   135°C   demonstrating   the   ability   to monitor loss of channel water from the hemipentahydrate.*

The second endotherm in the DTA response (Fig. 3) begins near 140°C and corresponds to loss of lattice water from the hemi-pentahydrate. Loss of crystal lattice structure during the dehydration could allow for spontaneous conversion to a more stable crystal form. If the recrystallization were to occur simultaneously with the dehydration process, the overlapping transitions (endothermic dehydration versus exothermic crystallization) can produce a DTA response in which the smaller energy event is difficult or impossible to detect. The corresponding Raman spectra collected between 100°C and 150°C are shown in Fig. 5. These spectra show the formation of a spectral peak at approximately 916 cm^{-1}. This peak is unique to the monohydrate form of Risedronate, providing evidence of monohydrate formation concurrent with loss of lattice water. The Raman spectrum collected at 150°C in Fig. 5 contains spectral features consistent with overdried hemi-pentahydrate, monohydrate, and anhydrate, indicating that both monohydrate and anhydrate are formed during dehydration. Spectral peaks unique to the anhydrate form are difficult to visually identify in Fig. 5 due to interference from the presence of monohydrate. However, outside of the spectral region shown in Fig. 5, peaks unique to anhydrate are observed. Quantitative estimates of phase composition may be possible from relative peak intensities of the Raman spectral peaks from each hydrate form, however, these calculations are not included in this report and are the subject of current spectral interpretation efforts.

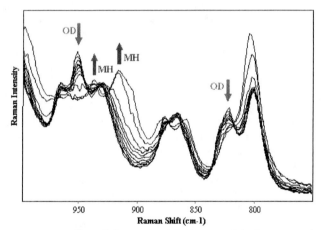

FIG. 5—*Raman spectra collected during the dehydration of the hemi-pentahydrate form of Risedronate between 100°C and 150°C showing the formation of monohydrate.*

The two highest temperature endotherms observed in the DTA (Fig. 3) correspond to removal of the last mole equivalent of water from the hemi-pentahydrate. The total amount of water removed from this material by 199°C supports the identification of Risedronate as a hemi-pentahydrate. This identification has been confirmed by X-ray crystallography. Raman spectra collected between 140°C and 215°C (Fig. 6) show the formation of a Raman spectral feature at approximately 910 cm^{-1} that is unique to the anhydrate form of Risedronate. Anhydrate formation between 150°C and 205°C indicates that a second recrystallization process occurs during the dehydration of this material beginning at approximately 160°C. However, evidence of this second recrystallization process is not observed in the DTA thermal profile shown in Fig.

3. At temperatures above approximately 215°C, thermal-induced degradation is observed in both the TG/DTA thermal profile and the corresponding Raman spectra.

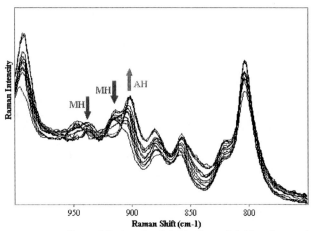

FIG. 6—*Raman spectra collected during the dehydration of the hemi-pentahydrate form of Risedronate between 140°C and 215°C showing the formation of anhydrate.*

To investigate potential relationships between channel water released during hemi-pentahydrate dehydration and subsequent recrystallization events, experiments similar to those summarized in Figs. 3–6 have been performed on overdried hemi-pentahydrate. In these experiments, most of the channel water in the crystal lattice was removed from the material prior to characterization by TG/DTA/Raman. The TG/DTA thermal profile observed during the dehydration of this overdried material is illustrated in Fig. 7. The TG/DTA profile for the overdried material is unusual in that an initial weight gain was observed for the high hygroscopic sample as it picked up water from the environment at the start of the scan. The dry nitrogen purge of 100 ml/min was insufficient to keep the sample chamber completely dry upon loading the sample. Results from these experiments are similar to those summarized in Figs. 3–6. Dehydration of the overdried hemi-pentahydrate initially formed a mixture of the starting material, monohydrate and anhydrate. At temperatures above approximately 150°C, spectral peaks unique to the anhydrate form are observed in the Raman spectra. At temperatures above approximately 215°C, evidence of thermal induced degradation is observed in both Raman spectra and the TG/DTA thermal profile. The temperature induced dehydration of overdried hemi-pentahydrate prepared by desiccation is similar in nature to dehydration of equilibrated hemi-pentahydrate subsequent to removal of channel water during the TG/DTA experiment. These results indicate channel water removed during the dehydration of hemi-pentahydrate do not influence subsequent recrystallization processes that occur during the dehydration of this material.

Similarly, potential relationships between lattice water released during hemi-pentahydrate dehydration and subsequent recrystallization events was investigated by conducting experiments similar to those summarized in Figs. 3–6 on the monohydrate form of Risedronate. The TG/DTA thermal profile observed during the dehydration of monohydrate is illustrated in Fig. 8. In this experiment, one mole equivalent of water is released from the material at temperatures above approximately 190°C. No evidence of recrystallization to produce the anhydrate form is observed in the DTA thermal profile.

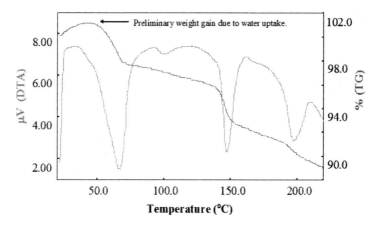

FIG. 7—*TG/DTA (5°C/min) thermal profile observed during the dehydration of overdried-hemi-pentahydrate form of Risedronate. (Exotherm up.)*

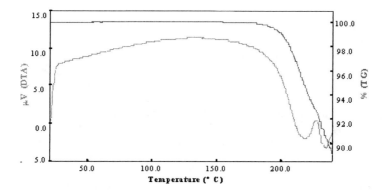

FIG. 8—*TG/DTA (5°C/min) thermal profile observed during the dehydration of the monohydrate form of Risedronate. (Exotherm up.)*

Raman spectra collected during the dehydration of the monohydrate are illustrated in Fig. 9. These spectra provide weak evidence for the formation of anhydrate during the dehydration of the monohydrate. Although peak shifting observed between approximately 915 cm^{-1} and 910 cm^{-1} is consistent with anhydrate formation, the Raman peak at approximately 800 cm^{-1} observed during monohydrate dehydration does not shift in a direction consistent with anhydrate formation. In addition, peaks unique to anhydrate outside the spectral region shown in Fig. 9 are not observed in the sample spectra. Therefore, results from these studies indicate anhydrate formation during dehydration of phase pure monohydrate is not observed.

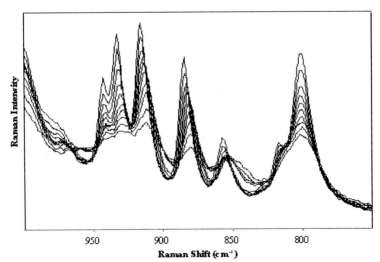

FIG. 9—*Raman spectra collected during the dehydration of the monohydrate form of Risedronate.*

Conclusions

This report has demonstrated benefits from coupling Raman spectroscopy to a commercial TG/DTA instrument to facilitate characterization of hydration states and dehydration processes. Raman spectra collected during the dehydration of Risedronate in these experiments provided valuable additional information for the characterization of these materials that was not readily available from the TG/DTA data. Change in lattice environment of the hemi-pentahydrate with loss of channel water, crystallization of the monohydrate form upon loss of lattice water from the hemi-pentahydrate, and finally crystallization of anhydrate upon dehydration of the monohydrate formed in situ were all events marked by the Raman data alone. The concurrent use of the Raman with the thermal instrument enabled clear definitions of the temperature ranges for the transitions of interest as well as the magnitude of the water loss associated with each transition. Clearly, the simultaneous collection of Raman spectra during the acquisition of a thermal profile offers advantages for the characterization of hydrated materials not realized by monitoring evolved gasses or conducting off-line analysis.

References

[1] Szekely, G., et al., *Thermochimica Acta*, Vol. 196, 1992, pp. 511–532.
[2] Kinoshita, R. et al, *Thermochimica Acta*, Vol. 222, 1993, pp. 45–52.
[3] Giron, D., *Journal of Thermal Analysis and Calorimetry*, Vol. 68, 2002, pp. 335–357.
[4] Vora, A. et al., *29th NATAS Conference Proceedings*, 2001, pp. 101–105.
[5] Loridant, S., Bonnat, M., and Siebert, E., *Applied Spectroscopy*, Vol. 49, 1995, pp. 1193–1196.
[6] Chang, H. and Huang, P. J., *Raman Spectroscopy*, Vol. 29, 1998, pp. 97–102.
[7] Murgan, R., Hhule, A., and Chang, H. J., *Applied Physics*, Vol. 86, 1999, p. 6779.

[8] Chang, H., et al., *Reviews in Analytical Chemistry*, Vol. 20, No. 3, 2001, pp. 207–238.

[9] Chang, H. and Huang, P. J., *Analytical Chemistry*, Vol. 69, 1997, pp. 1485–1491.

[10] Huang, P. J., et al., *Thermochimica Acta*, Vol. 297, 1997, pp. 85–92.

[11] Ghule, A., et al., *Thermochimica Acta*, Vol. 371, 2001, pp. 127–135.

[12] Chang, H., et al., *Materials Chemistry and Physics*, Vol. 58, 1999, pp. 12–19.

[13] Ghule, A., et al., *Inorganic Chemistry*, Vol. 40, 2001, pp. 5917–5923.

[14] Baskaran, N. and Chang, H., *Materials Chemistry and Physics*, Vol. 77, 2003, pp. 889–894.

[15] Murugan, R., et al., *Journal of Physics: Condensed Matter*, Vol. 12, 2000, pp. 677–700.

[16] Ghule, A., et al., *Spectrochimica Acta*, Vol. Part A 59, 2003, pp. 1529–1539.

[17] Chang, H., et al., *Thermochimica Acta*, Vol. 374, 2001, pp. 45–49.

[18] Murugan, R., et al., *Thermochimica Acta*, Vol. 346, 2000, pp. 83–90.

[19] Dubois, C. D., et al., 30th *NATAS Conference Proceedings*, 2002, pp. 214–219.

[20] Ghule, A. V., Lo, B., Tzing, S. H., Hhule, K., Chang, H., and Ling, Y. C., *Chem. Phys. Letters*, Vol. 381, 2003, pp. 262–270.

[21] Jager, H. J. and Prinsloo, L. C., *Thermochimica Acta*, Vol. 376, 2001, pp. 187–196.

[22] Ghule, A., Baskaran, N., Murugan, R., Chang, H., *Solid State Ionics*, Vol. 161, 2003, pp. 291–299.

[23] Redman-Furey, N. L., et al., 30th *NATAS Conference Proceedings*, 2002, pp. 733–738.

[24] *Analytical Applications of Raman Spectroscopy*, M. J. Pelletier, Ed., 1999, pp. 1–97.

[25] Wang, F., et al, *Organic Process Research & Development*, Vol. 4, 2000, pp. 391–395.

[26] Collins, W.J., et al., "Development and Evaluation of a TG/DTA/Raman System," *31st NATAS Conference Proceedings*, 2003.

[27] Redman-Furey, N. L., et al., *Journal of Pharmaceutical Sciences*, Vol. 94, No. 4, 2005, pp. 893–911.

Journal of ASTM International, October 2005, Vol. 2, No. 9
Paper ID JAI12794
Available online at www.astm.org

Carlton G. Slough, Ph.D.[1]

Quantitative Mass Measurements from Mass Spectrometer Trend Data in a TG/MS System

ABSTRACT: The use of mass spectrometry (MS) in the identification of evolved gaseous species in thermogravimetric (TG) experiments is well established. The mass spectrometer can be attached directly to the outgas port of the thermogravimetric instrument. Mass spectrometers have the ability to detect species to 1 ppm or better, and the detection is done in real time as the TGA scans. Typically, the mass spectrometer is used for identification only. However, there are situations where it would be desirable to quantify the MS data. For example, when two or more species are evolved during a weight transition, the total weight loss is known from the TGA, but the weights for the individual components are unknown. Calibration techniques for quantifying MS data are already well established. Following one of these techniques, we show that trend data from a mass spectrometer can be used to approximate the amount of material evolved for low m/e ratio species such as water and carbon dioxide. The process involves first calibrating the system using a sample that only gives off the gaseous species of interest. Multiple runs at different initial weights provide the data for the calibration. A correlation between mass loss as measured by the TGA and ion current increase as measured by the MS is then constructed. In this study, the technique is applied not only to overlapping transitions but also to the quantification of reaction products from gases released during a TGA experiment. An attempt is also made to quantify the accuracy of the technique.

KEYWORDS: Thermogravimetry, Mass Spectrometry, Compositional Analysis

Introduction

It is very common in the field of thermogravimetry to analyze the gases evolved during experiments by techniques such as mass spectrometry (MS) or Fourier transform infared spectrometry (FTIR) [1]. Such techniques are typically used for detection and identification purposes only. In the field of MS, it is quite common to go beyond identification and to analyze quantitatively complex gas mixtures in terms of percentages of various components [2]. It would be beneficial if quantitative information could be derived from TG/MS systems too. For example, when multiple species are released during a single weight loss event in a TGA, only the total weight loss is known. It would be useful to know how the weight loss is distributed among the various components. Also, if reactions occur between the evolved gases and/or the purge gas, information concerning the masses involved could be valuable.

If a mass spectrometer is attached to the TGA, then it is possible to analyze the ion current data for a specific m/e species during a weight transition and to convert it to a mass for that species. Numerous authors have explored the area of quantitative calibration of MS data in combined thermal/MS systems [3–6]. One technique explored by these authors and used here involves first calibrating the TG/MS system by running a sample that evolves only the gas species of interest. For example, if water is the component of interest, then calibration involves collecting TG/MS data from a sample that only gives off water during a weight transition. We show how the technique can be applied not only to the release, but also to the production of

Manuscript received 2 June 2004; accepted for publication 31 July 2005; published October 2005. Presented at ASTM Symposium on Techniques in Thermal Analysis:Hyphenated Techniques Thermal Analysis of the Surface and Fast Rate Analysis on 24-25 May 2004 in West Conshohocken, PA.
[1] Applications Chemist, TA Instruments, 109 Lukens Dr., New Castle, DE 19720.

gaseous species that occurs if reactions are occurring between evolved and/or purge gases in the TGA. The utility of the technique as applied to overlapping transitions is explored. Finally, it is shown that the accuracy of the method is within 25 %. Points concerning good experimental procedure for proper calibration are discussed.

Experimental

A TA Instruments Q50 TGA was attached to a Pfeiffer Vacuum ThermoStar mass spectrometer. The Q50 TGA has a quartz-lined furnace that is ideal for evolved gas experiments. Evolved gas components from the TGA are fed via a capillary heated to 200°C to the mass spectrometer. The input gas is ionized by electron impact. The ions are then sent through a quadrupole mass filter and finally impact onto a secondary electron multiplier (SEM) that produces an output current proportional to the ion current.

Powder samples of calcium oxalate monohydrate ($CaC_2O_4 \bullet H_2O$) from Fisher Scientific, calcium carbonate ($CaCO_3$, 99.97 %) from Alfa, calcium sulfate dihydrate ($CaSO_4 \bullet 2H_2O$, 99 %), and metoprolol tartrate ($C_{15}H_{25}NO_3 \bullet 1/2C_4H_6O_6$, 99 %) from Sigma Aldrich, were run in 100 μl platinum pans. The pans were cleaned and tared prior to sample loading.

Results and Discussion

Figure 1 shows a typical thermogravimetry scan of $CaC_2O_4 \bullet H_2O$ in a helium atmosphere. As shown below, $CaC_2O_4 \bullet H_2O$ decomposes in an inert atmosphere in three steps [2]. The decomposition steps are:

1. $CaC_2O_4 \bullet H_2O \leftrightarrow CaC_2O_4 + H_2O$ (loss of water)
2. $CaC_2O_4 \rightarrow CaCO_3 + CO$ (loss of carbon monoxide)
3. $CaCO_3 \leftrightarrow CaO + CO_2$ (loss of carbon dioxide)
If oxygen is present, a reaction with the CO produced in step 2 occurs:
4. $CO + \frac{1}{2} O_2 \rightarrow CO_2 + O$

Thus, in the presence of low amounts of oxygen, both CO and CO_2 are produced simultaneously during the second weight loss transition. Figure 1 also shows the mass spectral trend data captured for the calcium oxalate run. In a trend scan only certain m/e species are tracked over time by the mass spectrometer. In this example, data on m/e species 16, 17, 18, 28, 32, and 44 are collected and plotted. The typical ions associated with these m/e numbers are O, HO, H_2O, CO (N_2), O_2, and CO_2 respectively. The data show that during the second weight loss both CO and CO_2 are detected. The data also show that the O_2 level drops during the second weight loss. This is a definitive indication that reaction 4 is proceeding and is responsible for the production of the detected CO_2. The total weight loss for the second transition from Fig. 1 is 1.128 mg. This is all from CO, but it would be interesting to know how much of the CO is converted to CO_2, and how much CO_2 is thus produced. In order to probe this question, the MS system was calibrated to measure weight of evolved CO_2.

$CaCO_3$ thermally decomposes according to step 3 outlined above. The presence of oxygen at small levels does not make a difference [7]. Three different masses of calcium carbonate were run in the TG/MS system. Figure 2 shows an overlay of TG and MS trend data from a typical run. One weight loss transition occurs, and CO_2 (m/e = 44) is released. In this trend scan, increases in CO (m/e = 28) and O (m/e = 16) are also detected, but these are expected fragments of CO_2. The area under the CO_2 curve can be related to the weight loss detected in the TGA. Figure 3 shows how the area of the CO_2 peak grows with increasing mass of sample, and Fig. 4 shows the plot of weight loss as seen in the TGA versus integrated area under the CO_2 curve. A linear relationship is seen to exist between the curve area and the weight loss. Results from a linear fit are also displayed on the graph. The correlation is high, and the y intercept is near 0, as should be the case.

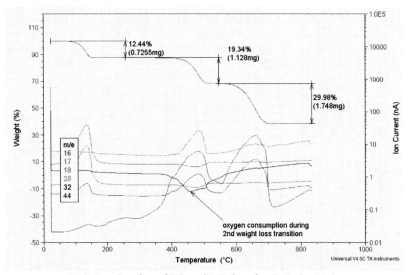

FIG. 1—Overlay of TG and MS data for $CaC_2O_4 \bullet H_2O$.

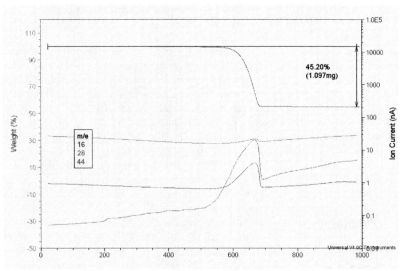

FIG. 2—Overlay of TG and MS data for $CaCO_3$.

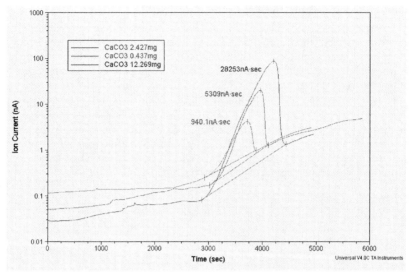

FIG. 3—*Integrated CO_2 peaks for increasing mass of $CaCO_3$.*

This curve can now be used to estimate the amount of CO_2 formed during the second transition in the $CaC_2O_4 \bullet H_2O$ data. Integration of the CO_2 peak from the second weight loss event in Fig. 1 gives an area of 39.58 nAmin. Inputting this into the linear equation obtained in Fig. 4 gives a calculated amount of 0.456 mg of CO_2 formed. Multiplying this by the ratio of the molecular weight of CO to the molecular weight of CO_2 gives a value of 0.290 mg of CO converted or 25.7 %.

Weight Loss vs. Carbon Dioxide Area

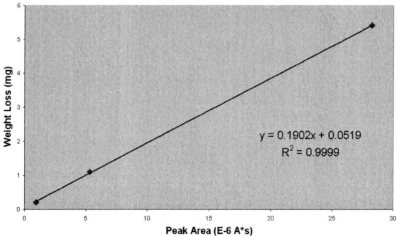

FIG. 4—*TG weight loss versus MS integrated area for CO_2.*

It is difficult to measure the accuracy of the technique. Strictly speaking, an alternate measure by a separate technique should be employed. However, as an approximate check of the accuracy of the method, the peak area of the CO_2 can be converted to a weight for the third weight loss event in calcium oxalate. Here only CO_2 is being released. Inserting the area under the CO_2 MS peak in to the equation above gives a value of 1.360 mg as compared to a value of 1.748 mg by the TGA. These values are within 22 % of each other.

Examples applying the technique to overlapping transitions are illustrated in Figs. 5–8. Figure 5 shows a typical TGA run on $CaSO_4 \bullet 2H_2O$. This compound has two waters of hydration. In order to resolve the release of these waters, as in this case, the sample is usually run in a hermetic pan with a pinhole through the lid. Figure 5 shows that the weight loss associated with the first water release is approximately 16 %, while that of the second is approximately 5 %. Multiple runs confirmed these as average percentages.

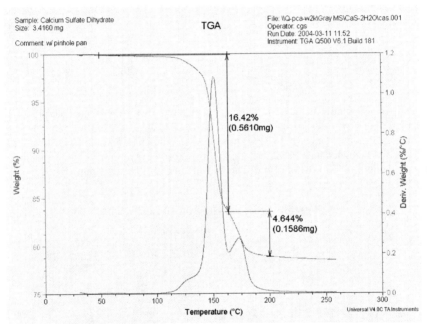

FIG. 5—*TGA plot showing water release in $CaSO_4 \bullet 2H_2O$. Use of a hermetic pan with a pinhole is used in order to resolve both waters of hydration.*

$C_{15}H_{25}NO_3 \bullet 1/2C_4H_6O_6$ is an organic drug known as a beta-blocker. It is used, among other things, in the treatment of hypertension. If mixed with $CaSO_4 \bullet 2H_2O$ and run in at TGA at 10°C/min, part of the decomposition of the $C_{15}H_{25}NO_3 \bullet 1/2C_4H_6O_6$ overlaps with the release of the second water of hydration from the $CaSO_4 \bullet 2H_2O$ and hinders proper resolution. This is shown in Fig. 6. In this experiment the $CaSO_4 \bullet 2H_2O$ is still run sealed in a hermetic pan with a pinhole. This pan is placed in the 100 ul platinum TGA pan as usual, and then the $C_{15}H_{25}NO_3 \bullet 1/2C_4H_6O_6$ is distributed around the outside, along bottom of the platinum pan. For this experiment 4.06 mg of $CaSO_4 \bullet 2H_2O$ was mixed with 4.94 mg of $C_{15}H_{25}NO_3 \bullet 1/2C_4H_6O_6$. If measurements are made of the water mass losses using the TGA graph, the results obtained are 0.758 mg for the first water and 0.498 mg for the second water. The measurement of the first water is reasonably accurate since relating to the initial mass of the $CaSO_4 \bullet 2H_2O$ gives 18.7 %. However, the measurement of the second is not, since relating it to the initial mass gives a loss of 12.3 %.

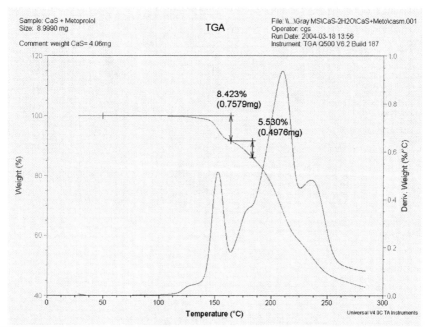

FIG. 6—*TGA plot of $CaSO_4 \bullet 2H_2O$ mixed with $C_{15}H_{25}NO_3 \bullet 1/2C_4H_6O_6$. The decomposition of the $C_{15}H_{25}NO_3 \bullet 1/2C_4H_6O_6$ obscures resolution of the second water loss and leads to an inaccurate measure.*

Figure 7 show a calibration curve for water obtained using the first weight loss in $CaC_2O_4 \bullet H_2O$. The plot is again very linear with a near zero intercept. Figure 8 plots the trend scan for m/e 18 for the data shown in Fig. 6. Note that the resolution of the water peaks is superior to that in the TGA graph. If this plot is integrated with perpendicular drops constructed as shown and the results plugged into the formula derived in Fig. 7, superior numbers are obtained. The result for the first water of hydration related to the initial mass of the $CaSO_4 \bullet 2H_2O$ gives 15.5 %. This is in good agreement with the measurement made on the pure material. For the second water of hydration, the result is 5.8 %, which is also in good agreement also with the TGA work done on the pure material and is a much-improved measurement over the TGA result from the mixture.

To ensure that the best accuracy is obtained, the following points concerning experimental procedure should be observed. The calibration must be done in the same purge gas as the sample being run. Changing purge gases could potentially shift baseline levels of the gases of interest. The calibration is sensitive to the baseline level, and if the baseline is not the same during the experimental run as during the calibration run, the results will be in error. Baseline shifts could also be related to changes and/or drift in the SEM voltage and, in the case of water, changes room in humidity. The error is roughly proportional to the ratio of the baseline levels.

Weight Loss vs. Water Peak Area

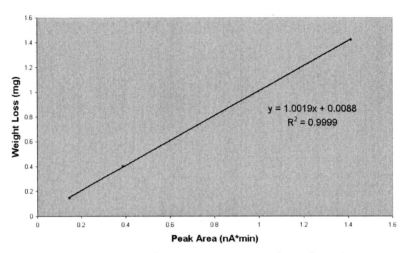

FIG. 7—*TG weight loss versus MS integrated area for H₂O.*

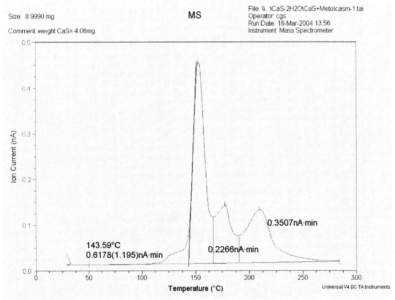

FIG. 8—*Plot of MS data for m/e =18 for the TG data plotted in Fig. 6.*

As a calibration material, powders are the best choice. Preliminary data show that use of chunk samples can produce a highly non-linear relation between peak area and weight loss.

Finally, an assumption being made here is that a gas fragments in the mass spectrometer always in the same proportions. For example, CO_2 will fragment into CO and O (and other fragments) always in the same proportions. Statistically, for large numbers of molecules, this assumption should be valid. Comparing the ratio of CO_2 to O for the three calibration runs can test this assumption. The ratio was the same in all three runs.

Conclusions

A simple calibration technique for converting areas under mass spectral data curves to weights of the detected species has been discussed. The technique was applied not only to examination of overlapping weight losses, but also to quantifying reaction products produced in the TGA. Results for low m/e species such as H_2O and CO_2 were presented with accuracies within 25 % obtained. While better accuracy is desirable, baseline shifts in the mass spectral data that occur make it difficult. The method gives reproducible results and is straightforward to perform. There is no reason why the technique could not be extended to higher m/e ratio ions.

References

[1] Turi, Edith A. (ed.), *Thermal Characterization of Polymeric Materials*, 2nd Edition, 1997, Vol. 1, pp. 49-73.

[2] Duval, Clement, *Inorganic Thermogravimetric Analysis*, 2nd Edition, 1963, pp. 281-282.

[3] Muller-Vonmoos, Max, Kahr, Gunter and Rub, Anton, *Thermochimica Acta*, 20, 1977, pp. 387-393.

[4] Price, D., Dollimore, D., Fatemi, N.S. and Whitehead, R., *Thermochimica Acta*, 42, 1980, pp. 323-332.

[5] Wang, Jisheng and McEnaney, Brian, *Thermochimica Acta*, 190, 1991, pp. 143-153.

[6] Maciejewski, M. and Baiker, A., *Thermochimica Acta*, 295, 1997, pp. 95-105.

[7] Wendlandt, Wesley Wm., *Thermal Methods of Analysis*, 2nd Edition, 1964, pp.16.

Journal of ASTM International, April 2005, Vol. 2, No. 4
Paper ID JAI12795
Available online at www.astm.org

Ramón Artiaga,[1] *Ricardo Cao,*[2] *Salvador Naya,*[1] *Bárbara González-Martín,*[3] *José L. Mier,*[1] *and Ana García*[1]

Separation of Overlapping Processes from TGA Data and Verification by EGA

ABSTRACT: A typical problem when analyzing the kinetics by dynamic TGA of polymer degradation consists in that different models can fit equally well, or the model does not fit the TGA data. It has been suggested that this is due to the presence of overlapping processes. High resolution TGA helps to clarify the problem in some cases, but in many other cases the problem remains unsolved. In this work, a dynamic TGA curve showing overlapping processes of degradation, typical of polymer thermal degradation, is fitted by a multiple logistic regression model. The regression function is a mixture of logistic functions where each single logistic function is weighed by a factor that represents the amount of the sample that takes place in the process represented by that function. A software program was developed to fit the data to the logistic mixture model by estimation of the parameter values. Additionally, the correlation of parameter estimates was performed by the S-Plus software. The regression was performed so that the best fitting shows the number of single decomposition processes that take place along the experiment, even if they are overlapping. Other parameters have the physical meaning of mass loss rate and position of the process in the time or temperature domain. The study of the correlations of parameter estimates gives a new insight on how each component fits a logistic model. The parameters obtained describe the kinetic of the processes in a form totally independent of the Arrhenius model. The separation of processes allows for analysis of each single process by classical methods. Although the model is mathematically supported, verification was done by comparing the results with the FTIR spectra obtained from the gases evolved along the experiment. The FTIR analysis of the gases evolved from a TGA experiment confirms that the logistic mixture model is suitable to analyze a dynamic TGA trace. It was also found that the logistic mixture model is suitable to separate single components from FTIR spectra. Some differences were found in the location of the components along the time axis, comparing DTG and FTIR data. In the FTIR, the separated peaks are broader and appear shifted to longer times. It can be attributed to the delay and mixing of components that may occur during the gas carriage from the TGA furnace to the FTIR detector.

KEYWORDS: fitting, TGA, logistic, overlapping, separation, smoothing

Introduction

Thermogravimetric Analysis (TGA) is frequently used to study polymer stability and degradation. In dynamic TGA, the mass of the sample is continuously monitored while the sample is subjected, in a controlled atmosphere, to a thermal program, where the temperature is ramped at a constant heating rate. The temperature of initial mass loss is usually viewed as the beginning of degradations [1]. Polymer degradation is a complex process that may involve combinations of random scission, depolymerization, and other mechanisms [2]. Constant heating rate TGA may be useful to identify and quantify components in a sample, but it is not possible

Manuscript received 22 May 2004; accepted for publication 30 September 2004; published April 2005. Presented at ASTM Symposium on Techniques in Thermal Analysis:Hyphenated Techniques Thermal Analysis of the Surface and Fast Rate A on 24-25 May 2004 in West Conshohocken, PA.

[1] Professor, Escola Politécnica Superior, University of Coruña, Mendizábal s/n 15403-Ferrol, Spain.
[2] Professor, Faculty of Computer Sciences, University of Coruña, Campus Elviña, A Coruña, Spain.
[3] Student, Faculty of Computer Sciences, University of Coruña, Campus Elviña, A Coruña, Spain.

when the decomposition of the components occurs simultaneously [3]. On the other hand, kinetic parameters of the process of degradation can be investigated by both isothermal and dynamic heating experiments [4].

Smoothing of TGA curves is generally performed when noise reduction is desired. It makes the data analysis easier, especially when derivatives calculation is concerned. For kinetic analysis purposes, sections of the TGA traces are normally fitted to different models.

It has been mentioned that the kinetic parameters obtained by fitting of single heating rate data to models is not suitable [5,6]. Multiple heating rate methods are considered more appropriate [7]. Different processes with different dependence on the heating rate may be overlapping. In those cases, the fitting to different models is usually not good.

A way to determine kinetic parameters for additively overlapping TG steps consists in fitting model equations to the non-standardized mass step by nonlinear optimization. The existence of an overlapping-free interval including the maximum for only one of the overlapping partial reaction steps was considered a condition for a successful separation of the peaks [8].

Concerning the smoothing, several methods were proposed in the literature [9–13].

A method to fit overall TGA traces was proposed by the authors [14]. The method overcomes the noise problem when analyzing TGA data and can be applied to overall TGA traces. It also overcomes the restriction consisting in the need of an overlapping-free interval including the maximum for one of the overlapping partial reaction steps.

The aim of this paper is to extend the application of the model to FTIR spectra obtained from the gases evolved from TGA experiments and to and compare the fitting with that obtained from the TGA data. It would allow for evaluation of the relative sensitivity and resolution of both hyphenated techniques to different components of the sample. In many cases, FTIR shows different sensitivity to different components that volatilize simultaneously in TGA experiments. In these cases, the observation of the FTIR spectra may confirm whether a logistic component obtained from a logistic mixture fitting corresponds to a real substance or not. On the other hand, the extension of the model to FTIR data will allow for separation of overlapping spectra, making their comparison with the IR spectra from databases easier.

TGA and FTIR Experiments

Samples of about 8 mg were cut off from oak timber and tested in a simultaneous thermogravimetric/differential thermal analyzer TA Instruments 2960 SDT. The sample was placed in aluminium crucibles and heated up to 590°C at 10°C/min, using a purge of 50 ml/min of N_2. Then, the gas was switched to air, and the temperature was held at 590°C for 20 min. The SDT instrument was attached to a Bruker Vector 22 spectrometer so that the FTIR spectra were continuously obtained from the gases evolved from the SDT. The ASTM E 2105 standard covers techniques that are of general use in the qualitative analysis of samples by TGA coupled with infrared spectrometric techniques [15].

Fitting the Data to a Logistic Mixture Model

Assuming that the degradation kinetics of each component of the sample can be represented by one or the sum of few functions, the trace versus time of a physical property, such as the sample mass, that is sensitive to thermal degradation, could be decomposed in several logistic functions. A TGA trace can be fitted by a combination of logistic functions:

$$Y(t) = \sum_{i=1}^{k} w_i f(a_i + b_i t)$$

$$f(t) = \frac{e^t}{1 + e^t}$$

(1)

where $i = 1, 2, \ldots, k$ represent different components in the model that in principle should be related to physical components in the sample.

In the case of absorbance of the evolved gases at a given infra-red frequency, a similar approach can be done to the TGA example, assuming that the integrated function of the absorbance gives a logistic-like curve.

The response $Y_i(t)$ should tend to asymptotic values when $t \to \infty$ and when $t = 0$. It implies that the b_i parameters have to be negative in case of TGA traces, where the sample mass typically decreases along time.

The function $Y_i(t)$ that represents the overall TGA or integrated FTIR trace may be expressed as a sum of $Y_i(t)$ functions like

$$Y_i(t) = w_i f(a_i + b_i t)$$

(2)

The constant b_i represents the slope of the i-th component (the largest rate of mass loss of the i-th component), while a_i / b_i is the value, in the time axis, where the i-th process is centered. The parameters w_i in Eq 2 account for the amount of mass explained by the i-th component.

Parameter Calculation

To estimate the parameter values in Eq 1, a software program was developed, although the correlations of parameter estimates were calculated by the nonlinear regression and related packages of the S-plus software. The following nonlinear regression model has been considered:

$$y_i = m(x_i, \theta) + \varepsilon_i, \quad i = 1, 2, \ldots, n$$

(3)

where the response variable and the independent variable values are denoted by y_i and x_i, respectively. θ is the parameters vector that is estimated by least squares, and ε_i are the errors, assumed to have normal distribution, zero mean, and constant variance.

The residuals of the model are defined as:

$$e_i(\theta) = y_i - m(x_i; \theta), \quad i = 1, 2, \ldots, n$$

(4)

The parameters of the model were estimated by the nonlinear least squares method, whose fundamentals were described by Gay [16].

The Levenberg-Marquart method for generation of the approximation sequence to the minimum point was used to calculate the parameter values that minimize that sum. It is based in the "trust region" algorithm, which was discussed by Chambers and Hastie [17]. Details about its implementation in S-plus are given in Dennis et al. [18]. A detailed study of the mixture of logistic functions was done by Naya [19].

Results and Dicussion

TGA and FTIR Raw Data

Figure 1 shows the data obtained from an oak sample in a combined TGA-FTIR experiment. The first mass loss step was attributed to moisture evaporation in this kind of material [20]. The second one corresponds to the pyrolysis of the sample and, looking at the DTG trace, seems to involve at least two important and one smaller overlapping processes. The FTIR spectra confirm this assumption, showing a double peak at several frequencies at the same time, in the time axis, as the DTG trace. The third mass loss step corresponds to the char combustion under air purge and is also accompanied by DTG and FTIR peaks. The peaks corresponding to the second mass loss step are evident at 2357, 1749, and 1174 cm^{-1}.

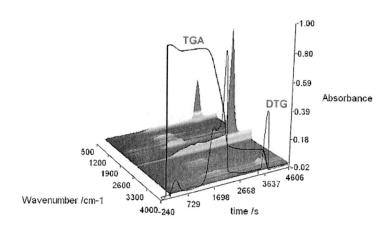

FIG. 1—*Quercus robur. The spectra obtained by FTIR analysis of the gases evolved from the TGA experiment. The TGA and DTG plots, arbitrarily scaled, were superimposed on the x axis.*

Figure 2 plots an overlay of the DTG along with the spectra obtained at 2357, 1773, 1745, and 1182 cm^{-1}. The 2357 cm^{-1} band is very sensitive to CO_2 and suggests that the combustion can happen to some extent even when the purge was nitrogen. The 1773 and 1182 cm^{-1} bands show two peaks that correlate very well with the second and third DTG peaks. A shifting of 179 s was measured between the DTG and FTIR peaks. It is important to notice that although in all the cases the peaks are overlapping, the DTG peaks seem more resolved. It may be due to the fact that the components of the gas may mix while they are carried through the transfer line from the TGA to the FTIR. Any improvement in this connection could result in more sharp peaks. Another important fact is that different frequencies exhibit different sensitivities to the overlapping components, which is useful to identify single components. Particularly, in the range from 1400–2600 s, the 1745 cm^{-1} band is much less sensitive to the first component than to the second one, while the 1773 cm^{-1} band reflects both components similarly.

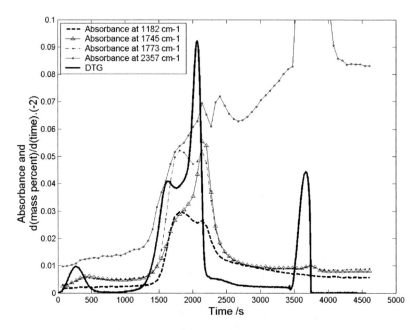

FIG. 2—*Absorbances, at different frequencies, obtained continuously from the evolved gases. For comparison of the peak locations on the time axis, a scaled DTG plot was superimposed.*

TGA Separation

Figure 3 shows the area under study for separation of the TGA curve in four components. The two DTG peaks and a small shoulder inform about three degradation processes. Since the TGA does not reach a zero value after the shoulder, a fourth underlying process may exist.

Figure 4 shows the four components obtained by the fitting to a mixture of logistics. The fitting obtained by the sum of the four components match the TGA data well, although a slight difference appears in some points. This deviation may be attributed to the fitting of the second component. Table 1 shows the correlations between the estimated weight parameters and their location and rate parameters. These inform about the relationship between the estimators of the weight parameters of every component and the location and scale parameters of that component. The closer the correlation is to 1 or −1, the higher the relation is. In the case of the second component, the correlation is slightly smaller than in the other cases. It suggests that a slightly different function could fit better for this component. Table 2 shows the parameter values obtained in the fitting and the point a_i/b_i where each process is "centered."

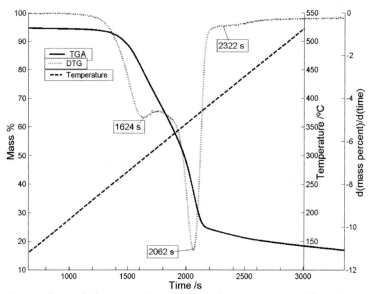

FIG. 3—*Area under study for separation of TGA in four components. The points stamped on the DTG trace were taken as starting points where three degradation processes are supposed to be centered.*

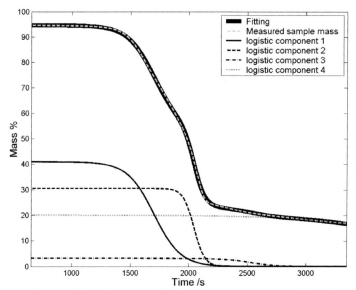

FIG. 4—*Overlay of the TGA trace, the fitting by a mixture of four logistics, and the separated components.*

TABLE 1—*Correlations between the estimated parameters of the mixture of logistics calculated for each component.*

	1st Component	2nd Component	3rd Component	4th Component
a_i and w_i	−0.906	−0.873	−0.928	−0.991
b_i and w_i	−0.919	−0.869	−0.933	−0.988

TABLE 2—*Parameter values obtained in the fitting and the point ai/bi where each process is "centered."*

	1st Component	2nd Component	3rd Component	4th Component
w_i	40.95840000	30.50780000	3.14710000	20.07400000
a_i	14.79970000	44.67980000	22.45620000	9.43530000
b_i	0.00869551	0.02185540	0.00885222	0.00233722
a_i/b_i	1702	2044	2537	4037

FTIR Separation

A similar approach to the TGA separation was applied to the absorbance data. In this case, the data were previously integrated in order to fit to a logistic model. Figures 5 and 6 show the integrated absorbance curves and the fitting and the single curves involved in the fitting corresponding to the 1182 and 1745 cm^{-1} bands, respectively. In principle, these frequencies are not very sensitive to CO_2, so this component should not appear in the fitting.

Figures 7 and 8 show the fitting of the differentiated function versus time and its components, overlayed on the measured spectra. Finally, Figs. 9–11 show the comparison of the two main components that were found in the TGA and FTIR fittings.

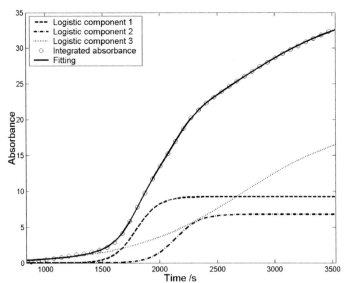

FIG. 5—*The integrated absorbance versus time curve, the fitting and the single curves involved in the fitting corresponding to the 1182 cm^{-1} band.*

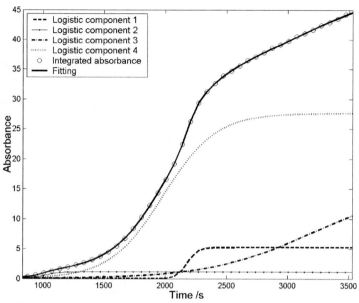

FIG. 6—*The integrated absorbance versus time curve, the fitting and the single curves involved in the fitting corresponding to the 1745 cm^{-1} band.*

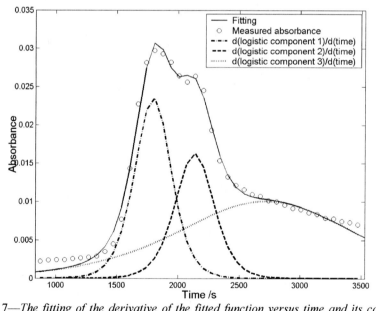

FIG. 7—*The fitting of the derivative of the fitted function versus time and its components, overlayed on the measured spectrum at 1182 cm^{-1}.*

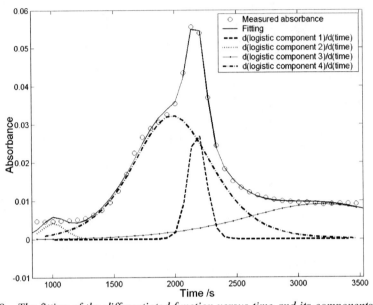

FIG. 8—*The fitting of the differentiated function versus time and its components, overlayed on the measured spectrum at 1745 cm^{-1}.*

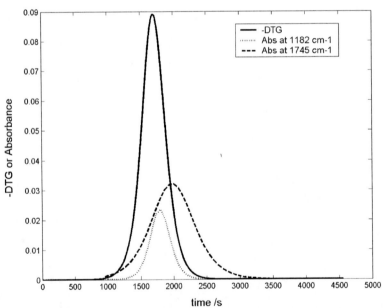

FIG. 9—*Comparison of the first main component found from TGA and from FTIR absorbance data at two different wave numbers.*

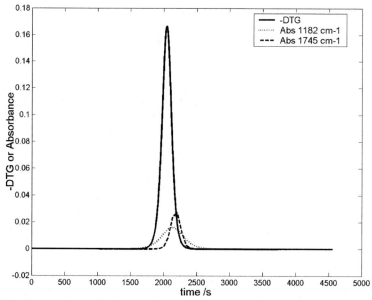

FIG. 10—*Comparison of the second main component found from TGA and from FTIR absorbance data at two different wave numbers.*

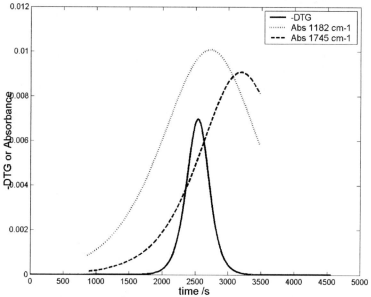

FIG. 11—*Comparison of the third main component found from TGA and from FTIR absorbance data at two different wave numbers.*

Conclusions

The FTIR analysis of the gases evolved from a TGA experiment confirms that the logistic mixture model is suitable to analyze a dynamic TGA trace. The logistic mixture model is suitable to separate single components from FTIR spectra.

Some differences were found in the location of the separated components along the time axis, comparing DTG and FTIR data. In the case of FTIR, the peaks appear shifted to longer times with respect to the DTG case. The FTIR peaks of the single components are broader than DTG peaks, considering the relation of the width with respect to the height. It can be attributed to the delay and mixing of components that may occur during the gas flow from the TGA furnace to the FTIR detector.

Acknowledgments

The authors gratefully acknowledge the grants BFM2002-00265 from MCyT (European ERDF support included) and PGIDIT03PXIC10505PN and PGIDIT04PXI105011F from Xunta de Galicia for the first three authors.

References

[1] Xie, W., Heltsley, R., Cai, X., Deng, F., Liu, J., Lee, C., et al., "Study of Stability of High–Temperature Polyimides Using TG/MS Technique," *Journal of Applied Polymer Science*, 83, 2002, pp. 1219–1227.

[2] Chartoff, R. P., In: Turi, E. A., Ed., "Thermal Characterization of Polymeric Materials," Vol. 1, 2nd ed., Academic Press, San Diego, 1997, p. 688.

[3] Bowley, B., Hutchinson, E. J., Gu, P., Zhang, M., Pan, W-P., and Nguyen, C., "The Determination of the Composition of Polymeric Composites Using TG–FTIR," *Thermochimica Acta*, 200, 1992, pp. 309–315.

[4] Sircar, A. K., In: Turi, E. A., Ed., "Thermal Characterization of Polymeric Materials," Vol. 1, 2nd ed., Academic Press, San Diego, 1997, p. 1254.

[5] Vyazovkin, S., "Two Types of Uncertainty in the Values of Activation Energy," *J. Therm. Anal. Calorim.*, 64, 2001, pp. 829 – 835.

[6] Vyazovkin, S. and Wight, C. A., "Model-Free and Model-Fitting Approaches to Kinetic Analysis of Isothermal and Nonisothermal Data," *Thermochim. Acta*, special issue honoring Prof. D. Dollimore, 340/341, 1999, pp. 53–68.

[7] Brown, M. E., Maciejewski, M., Vyazovkin, S., Nomen, R., Sempere, J., Burnham, A et al., "Computational Aspects of Kinetic Analysis. Part A: The ICTAC Kinetics, Project: Data, Methods, and Results," *Thermochim. Acta*, special Millennium issue, 2000, 355, pp. 125–143.

[8] Boy, S. and Boehme, K., "Kinetic Analysis of Additively Overlapping Reactions. Part 1. Description of an Optimization Method and Its Use for the Separation of Peaks," *Thermochimica Acta*, 75, 1984, pp. 263–73.

[9] Wang, K., Wang, S., Huang, H., Klein, M. T., and Calkins, W. H., "A Novel Smoothing Routine for the Data Processing in Thermogravimetric Analysis," Preprints of Papers – American Chemical Society, Division of Fuel Chemistry, 41, 1996, pp. 27–32.

[10] Whittem, R. N., Stuart, W. I., and Levy, J. H., "Smoothing and Differentiation of Thermogravimetric Data by Digital Filters," *Thermochimica Acta*, 57, 1982, pp. 235–239.

[11] Varhegyi, G. and Till, F., "Computer Processing of Thermogravimetric-Mass Spectrometric and High Pressure Thermogravimetric Data. Part 1. Smoothing and Differentiation," *Thermochimica Acta*, 329, 1999, pp. 141–145.

[12] Naya, S., Cao, R., and Artiaga, R., "Local Polynomial Estimation of TGA Derivatives Using Logistic Regression for Pilot Bandwidth Selection," *Thermochimica Acta*, 406, 2003, pp. 177–183.

[13] Artiaga, R., Cao, R., Naya, S., and Trillo, J., "Nonparametric Two-Stage Plug-In Adpative Smoothing for Thermal Analysis Data," *Journal of Statistical Computation and Simulation*, in press.

[14] Cao, R., Naya, S., Artiaga, R., García, A., and Varela, A., "Logistic Approach to Polymer Degradation in Dynamic TGA," *Polymer Degradation and Stability*, in press.

[15] ASTM Standard E 2105, "Standard Practice for General Techniques of Thermogravimetric Analysis (TGA) Coupled with Infrared Analysis (TGA/IR)," ASTM International, West Conshohocken, PA.

[16] Gay, D. M., Lect. Notes Math, 1066, 1984, pp. 72–105.

[17] Chambers, J. M., and Hastie, T. J., Eds., "Statistical Models in S," Chapman & Hall, New York, 1991.

[18] Dennis, J. E., Gay, D. M., and Welsch, R. E., "An Adaptive Nonlinear Least–Squares Algorithm," ACM Transactions on Mathematical Software 7, 1981, pp. 348–368.

[19] Naya, S., "Nuevas Aplicaciones de la Estimación Paramétrica y No Paramétrica de Curvas al Análisis Térmico," Ph.D. Thesis, A Coruña Spain, Universidade da Coruña, 2003.

[20] Wiedemann, H. G. and Lamprecht, I., "Wood," In: *Handbook of Thermal Analysis and Calorimetry, Vol. 4: From Macromolecules to Man*, R. B. Kemp, Ed., Amsterdam, Elsevier, 1999, p. 793.

Journal of ASTM International, Vol. 3, No. 9
Paper ID JAI100351
Available online at www.astm.org

Quentin Lineberry,[1] *Thandi Buthelezi,*[1] *and Wei-Ping Pan*[1]

Characterization of Modified Carbon Nanotubes by TG-MS and Pyrolysis-GC/MS

ABSTRACT: Two evolved gas analysis techniques were used to study modified single-wall carbon nanotubes (SWNTs). These techniques were thermogravimetry-mass spectrometry (TG-MS) and pyrolysis-gas chromatography/mass spectrometry (pyrolysis-GC/MS). Increased weight loss was exhibited in the modified SWNTs below 800°C, which was attributed to the addition of methoxide introduced during the modification. The localized environments of the SWNT provide bonding sites of varying strength that results in the staged evolution of the methoxide masses. The continued evolution of methoxide related masses is a result of the variations in the localized environments of the SWNTs, including the curved ends and various other possible sites along the sidewalls. However, once all the modifications have been removed, the overall structure of the SWNT still remains as intact as the pristine sample as indicated by the slope of the weight loss at higher temperatures.

KEYWORDS: carbon nanotube, TG-MS, pyrolysis-GC/MS

List of Abbreviations

CNT = Carbon nanotube
ESC = Electrostatic charge mitigation
ODCB = *ortho*-dichlorobenzene
SDT = Simultaneous DSC-TGA
TG-MS = Thermogravimetry-mass spectrometry
SWNT = Single-wall carbon nanotube
NaOMe = Sodium methoxide
DTG = Derivative thermogravimetry
HiPco = High-pressure carbon monoxide
GC/MS = Gas chromatography/Mass spectrometry

Introduction

A number of applications have been envisioned for carbon nanotubes (CNTs) since their discovery. They have been studied for use in potential applications such as gas storage media, electronics, AFM tips, solar collectors, and reinforcement for nanocomposite materials [1–3]. In addition to perhaps enhancing the mechanical properties of certain nanocomposites, the addition of SWNTs can also impart electrical conductivity. For certain aerospace applications, such as certain components on Gossamer spacecraft, the electrical conductivity provided by the CNTs is sufficient to mitigate electrostatic charge (ESC) buildup on polymers [4]. A representation of a SWNT can be seen in Fig. 1. One of the current hindrances for widespread use of CNTs in projects, such as Gossamer spacecraft, is their inability to be evenly dispersed either in a polymer matrix or in a solvent. Several groups have attempted to overcome this problem by a

Manuscript received January 31, 2006; accepted for publication June 30, 2006; published online October 2006. Presented at ASTM Symposium on Techniques in Thermal Analysis: Hyphenated Techniques, Thermal Analysis of the Surface, and Fast Rate Analysis on 24 May 2004 in West Conshohocken, PA; L. Judovits and W.-P. Pan, Guest Editors.
[1] Graduate Research Assistant and Professors, respectively, Western Kentucky University, Chemistry Department, Bowling Green, KY 42101-1079.

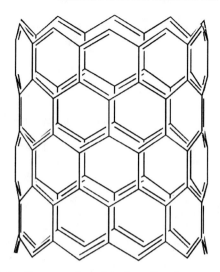

FIG. 1—*Representation of single-wall carbon nanotube.*

variety of means, including covalent modification, noncovalent modification, and sonication [5–7]. Covalently attaching modifying agents to the surface of the CNT is most often accomplished by reacting groups with the carboxylic acid groups introduced during purification. Noncovalent modifications can take advantage of the same carboxylic acid groups, but an ionic interaction between the carboxyl group and usually the amino group of a surfactant is formed instead.

Investigators at NASA Langley and the National Institute of Aerospace Research have obtained CNT samples that are well dispersed in select solvents, including tetrahydrofuran (THF), dimethylacetamide (DMAc), and 1,2-dichlorobenzene (ODCB). Following cited fullerene chemistry, the SWNTs were reacted to produce SWNTs with enhanced dispersibility as shown in their ability to stay dispersed in solvent for days as opposed to minutes. However, upon further examination, it was determined that the intended functionalization had not occurred, and some other phenomenon was responsible for the CNTs increased dispersibility. By eliminating reagents, it was determined that only sodium methoxide and pyridine were needed in the reaction mixture to produce SWNTs with enhanced dispersion. The purpose of this study was to determine the nature and extent of the modification that was responsible for the enhanced dispersion of the SWNTs. This was done by determining the thermal properties and evolution compounds in pyrolysis experiments.

Experimental

HiPco® (high-pressure carbon monoxide) SWNTs were purchased from Carbon Nanotechnologies, Inc. They were purified by heating at 250°C for 16 h in a high humidity chamber followed by Soxhlet extraction in hydrochloric acid (~22.2 weight %) for 24 h. Pyridine and sodium methoxide (NaOMe) were purchased from Aldrich Chemical Company and used as received. Methanol and 1,2-dichlorobenzene (ODCB) were obtained in reagent grade from Aldrich and used as received.

Four SWNT samples were studied. The purified SWNTs are denoted as A. Sample B was sonicated for 3 h in ODCB followed by heating to 80°C for 16 h while stirring in a nitrogen atmosphere. The SWNTs were then washed three times in methanol and collected via centrifugation. The collected sample was dried under vacuum at 100°C for 3 h. Sample C was sonicated for 3 h in pyridine followed by addition of NaOMe. The mixture was heated to 80°C for 16 h while stirring in a nitrogen atmosphere and then heated to reflux for 3 h. The pyridine was removed by distillation followed by collection of the nanotubes via centrifugation. The centrifuged sample was washed three times with methanol and dried under vacuum at 100°C overnight. The SWNT sample labeled D was sonicated in ODCB for 3 h and then added dropwise to a flask containing NaOMe and pyridine that were vigorously stirred. This mixture was heated to 80°C

TABLE 1—*Summary of experimental conditions.*

Sample	Sonicated for 3 h in:	Heated for 16 h in:
A	...	...
B	ODCB	ODCB
C	Pyridine	Pyridine /NaOMe
D	ODCB	ODCB / Pyridine / NaOMe

for 16 h while stirring in a nitrogen atmosphere, and then heated to reflux for 3 h. The pyridine was removed by distillation followed by collection of the nanotubes via centrifugation. The centrifuged sample was washed three times with methanol and dried under vacuum at 100°C overnight. The experimental conditions are summarized in Table 1.

The TG-MS system consists of a TA Instruments 2960 Simultaneous DSC-TGA (SDT) (TA Instruments, New Castle, DE) interfaced with a Fisons VG Thermolab Quadrupole Mass Spectrometer (VG Gas Analysis Systems, Cheshire, England) using a heated capillary transfer line. The capillary transfer line was heated to 170°C, and the inlet port of the mass spectrometer was heated to 150°C. The capillary inlet is constructed from fused silica encased within a stainless steel sheath. This is further covered with PTFE and red fiberglass sheathing. The probe is made of alumina ceramic. The system is operated at 70 eV and a pressure of 1×10^{-6} torr with a detection range of $1-300$ amu.

Pyrolysis-GC/MS allows for a more definitive identity determination to be made. However, unlike TG-MS, relating a certain species to a specific mass loss event is not as straightforward. Pyrolysis-GC/MS works by cryogenically trapping all compounds evolved over a given temperature range and then introducing all the trapped compounds to the GC as one small sample plug by rapidly heating the cryogenic focusing system. Beyond sample introduction, the technique behaves just like a standard GC-MS experiment with the compounds being separated by the GC, and then identified using the MS. A given species can be related to a given mass loss event only if the sample is pyrolyzed over one mass loss event.

The Pyrolysis/GC-MS system consists of a LECO® ThermEx Inlet System, a LECO® Pegasus® II GC/MS system (LECO Corporation, St. Joseph, MI), and a Smart Cryo™ Smart Cryogenic Focusing Device (HISI Analytical Technologies, Humble, TX). The Pegasus® II GC/MS system includes a HP 6890 High Speed Gas Chromatograph (Agilent Technologies, Palo Alto, CA) and a time-of-flight (TOF) mass spectrometer (LECO Corporation, St. Joseph, MI). The advantage of TOF mass spectrometry is its potential for tremendously fast acquisition rates. The ThermEx Inlet System is designed to heat small quantities of a sample in a quartz pyrocell and transfer the volatilized sample components to a heated capillary GC injection port. The carrier gas connections and seals have been designed to provide a leak-free system for operation with GC-MS system. The associated GC has a capillary injector capable of operating in the split mode and an oven configured with a cryogenic option (liquid nitrogen). During the sample heating period, evolved volatiles are condensed at the column head by maintaining the GC oven at cryogenic temperature. Once the sample has been heated to the desired temperature in the ThermEx™ system and all out-gassed components condensed at the column head, the cryogenic system is very rapidly heated to vaporize the components and inject them into the GC column for analysis. The GC capillary column used was HP-5 with a size of 30 m by 0.32 mm by 0.25 μm. The GC oven program was set as follows: ramp from 50 to 300°C with a heating rate of 50°C/min. The injection temperature was 295°C.

Results and Discussion

SDT

Figure 2 shows the TGA curves for the four samples. There are two main mechanisms at work in the temperature range studied—below 1000°C and above 1000°C. Below 1000°C, degradation and devolatilization predominate. Above 1000°C, oxidation predominates, which is supported by the MS data that will be discussed later. From the TGA curves, the similar weight loss rates ($\sim$0.4 %/°C) and the observed CO_2 evolution, which will be discussed in the MS results, indicate the presence of less than 0.001 % oxygen in the UHP nitrogen purge gas. The only way to avoid this small oxygen content would be to use a vacuum system, which would interfere with the operation of the mass spectrometer. The main focus of this study was the low temperature region below 800°C.

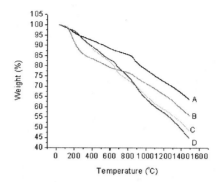

FIG. 2—*TGA plots of all samples.*

In the low temperature devolatilization range, the pristine SWNT sample, Sample A, has several weight losses, but three appear to dominate not including the initial water loss around 100°C (Fig. 3). There is a DTG peak maximum at approximately 170°C, which may be due to the loss of small molecules adsorbed during their prior treatment. Following that evolution is a less well-defined weight loss that occurs from 350 to 750°C. Finally, there is another DTG peak at 865°C. The DTG curve for Sample B is very similar to sample the DTG curve for Sample A except that the intensity of the 170°C DTG peak has increased. The difference in residues remaining (63.6 and 55.54 %, respectively) at the end of the heating program is roughly equivalent to the difference in mass loss during the 170°C mass loss, possibly implying adsorption of a reagent. This increased weight loss around 170°C for Sample B will be clearly explained in the MS data.

Initially, the soluble samples, C and D, lose less weight below 300°C. This reduction in initial weight loss results from the removal of unwanted amorphous material during the reaction and recovery processes. After 300°C, CNTs C and D begin to decompose faster, as indicated by the slope of their curves (Fig. 2). After 1000°C, the weight loss slopes of all the samples become approximately equal indicating that the degradation is occurring at the same rate. This indicates that once all the modified portions and adsorbents have been removed, the samples are essentially the same.

Samples C and D (Fig. 4) have several more weight losses when compared to Samples A and B. The DTG curve for Sample C differs from Sample D due to the addition of a peak at 410°C, and the shifting of the high temperature peak by roughly 100°C. The weight loss maxima occur at roughly 100, 140, 275, 410, and 700°C. The first two are likely due to moisture. The remainder of the weight losses, up to approximately 800°C, are likely due to the addition of the methoxide ion in the reaction mixture. The different decomposition stages seen between 200 and 800°C in both Samples C and D may be due to the

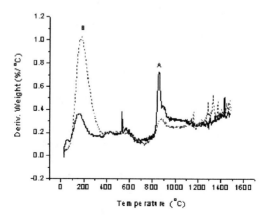

FIG. 3—*DTG curves of Samples A and B.*

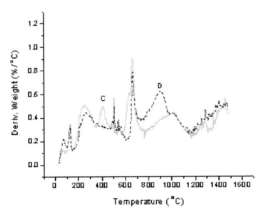

FIG. 4—*DTG curves of Samples C and D.*

different localized environments of the CNT surface—tube ends and defects along the tube wall, which would account for the multiple maxima seen in the DTG curves. Also, the multiple peaks could be the result of different types of nanotubes—ones with amorphous carbon and ones without amorphous carbon [8]. The amorphous carbon could have been distributed onto the tubes as a result of the sonication treatment to initially disperse the CNTs. The MS data support this assertion by showing the evolution of several similar mass numbers (14, 15, 16, 28, and others that are not shown) below 800°C as can be seen in Figs. 7 and 8.

Mass Spectrometry

Figure 5 shows the evolution of mass 44 from the various CNTs in a nitrogen atmosphere. As can be seen from this figure, all the samples have a distinct peak that is followed by continuing evolution. At this point it is worth noting that the total CO_2 evolution was relatively small as indicated by the range of intensity covered. The CO_2 peaks correspond well to the DTG curves. Samples A and B have peak CO_2 evolution at 885 and 874°C, respectively. Evolution of CO_2 from A and B are almost identical. The only difference between the two samples is the adsorption and subsequent evolution of ODCB from Sample B, which will be discussed shortly. Samples C and D have peak evolution at 684 and 689°C, respectively. Their CO_2 curves are similar but the curve for Sample C diverges from that of Sample D at roughly 600°C. This is likely due to the additional DTG peak of Sample C just prior to this evolution, which indicates another local environment of modification.

Certain species, such as the solvents and reactants used, can be expected in the MS results. Dichlorobenzene appears in Samples B and D, but not in Samples A and C, since it was not used. Evolution of the top four masses for dichlorobenzene from Sample B, as based on the NIST webbook, can be seen in

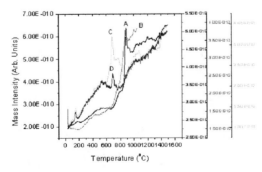

FIG. 5—*Evolution of m/z 44 as a function of temperature for all samples.*

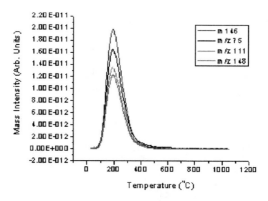

FIG. 6—*Evolution of the four most intense masses of dichlorobenzene from Sample B.*

Fig. 6. ODCB evolves during the first major weight loss from about 100 to 400°C with a peak at 190°C, which corresponds well with the boiling point of ODCB (~180°C). This implies that the ODCB had adsorbed onto the surface of the CNT. Evolution of dichlorobenzene is what causes the second mass loss event of CNT B to be so much bigger than that of CNT A. Given that the evolution of carbon species from CNTs A and B are almost identical, the only difference between the two is that CNT B absorbed some of the solvent, ODCB. The other species evolving at this temperature, which was discussed in the previous section and tentatively attributed to small molecules adsorbed during prior treatment, is most likely an alkane species as evidenced by the masses evolved and confirmed by infrared spectroscopy data.

Pyridine could also be expected in some of the samples, namely C and D. Evolution of this species can be seen, but the peak is not as well defined as the dichlorobenzene peaks discussed earlier. Also, evolution of pyridine occurs at a higher temperature than that of dichlorobenzene from around 200 to 700°C.

When the masses related to methanol/methoxide were examined, an interesting evolution was observed. Figures 7 and 8 show the evolution of select methoxide masses from Samples C and D. This evolution occurs over a wide temperature range from the initial temperature to around 800°C, which is unusual given the normally more well defined evolutions seen in TG-MS. Based on this, covalently attached methoxide in a range of localized environments could be imparting the enhanced dispersion to the SWNTs. The methoxide could have esterified the carboxylic acid groups present on the surface of the SWNTs or reacted with other defect structures on the SWNTs to alter the surface chemistry of the SWNT to allow for better dispersion, or both.

Pyrolysis-GC/MS

There are several compounds identified with the MS NIST database of the pyrolysis-GC/MS that all of the SWNT samples produce—benzene, toluene, ethylbenzene, p-xylene, 1,2,3-trimethyl benzene, 1-ethyl,

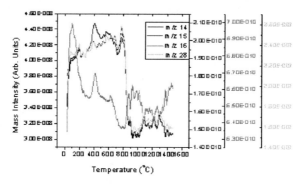

FIG. 7—*Evolution from Sample C of m/z values 14, 15, 16, and 28.*

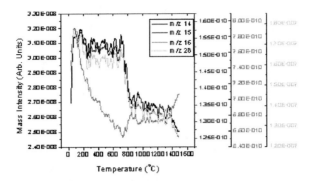

FIG. 8—*Evolution from Sample D of m/z values 14, 15, 16, and 28.*

3-methyl benzene, naphthalene, and biphenyl. The compounds can be assumed to be a result of decomposition of SWNT-like materials. The presence of benzene and its derivatives in the evolved gas is expected given that the starting material is a rolled graphite sheet.

However, the following compounds only evolved from the soluble samples, C and D: methanol, pyridine, methyl pyridines, phenol, 1,3-cyclohexadiene, 2-ethyl hexanal, and cis-a-methylstyrene. Pyridine should be expected in the soluble samples, given the reactants used for the functionalization. The other compounds could be a result of unforeseen degradation schemes involving the CNTs, amorphous carbon, and the substituent providing enhanced dispersion. The detection of these compounds in addition to the ones found by TG-MS is only indicative of the increased sensitivity of the GC-MS compared to the TG-MS.

The pyrolysis-GC/MS results indicate that methanol is one of the compounds only evolved from the soluble samples. Although methoxide was added to the reaction mixture, the GC/MS database said that the closest match was methanol. This lends some credence to the theory posed in the TG-MS section that methoxide is the species responsible for enhancing the dispersibility of the SWNTs. Methanol was used to wash the samples three times prior to analysis, but this does not explain why methanol was only detected in the soluble samples. Therefore, addition of methoxide to the surface of the CNT must take place prior to washing, and it must be impervious to the washing. This suggests that the functionalization has a more covalent nature; otherwise, the washing process would have likely removed it.

Conclusions

When enhanced dispersibility is achieved, the weight loss below 800°C increases due primarily to the addition of methoxide. The localized environments of the CNT provide bonding sites of varying strength, which results in the staged evolution of the methoxide masses. However, once all the modifications have been removed, the structure of the CNT still remains as intact as the pristine sample as indicated by the slope of the weight loss at higher temperatures. The continued evolution is a result of the varying localized environments of the SWNTs, including the curved ends and various other possible sites along the sidewalls. ODCB adsorbed during the reaction process evolved neatly with a peak intensity corresponding to the boiling point of ODCB.

Acknowledgments

The authors gratefully acknowledge the assistance and cooperation of John Connell and Donavon Delozier of NASA Langley Research Center and the financial assistance from the Center through GSRP grant NGT-01-03012.

References

[1] Bom, D., Andrews, R., Jacques, D., Anthony, J., Chen, B., Meier, M., and Selegue, J., "Thermogravimetric Analysis of the Oxidation of Multiwalled Carbon Nanotubes: Evidence for the Role of Defect Sites in Carbon Nanotube Chemistry," *Nano Lett.*, Vol. 2, No. 6, 2002, pp. 615–619.

[2] Cao, A., Xu, C., Liang, J., Wu, D., and Wei, B., "X-Ray Diffraction Characterization on the Alignment Degree of Carbon Nanotubes," *Chem. Phys. Lett.*, Vol. 344, No. 1–2, 2001, pp. 13–17.

[3] Salvetat-Delmotte, J.-P. and Rubio, A., "Mechanical Properties of Carbon Nanotubes: A Fiber Digest for Beginners," *Carbon*, Vol. 40, No. 10, 2002, pp. 1729–1734.

[4] Smith, J. G., Jr., Delozier, D. M., Connell, J. W., and Watson, K. A., "Carbon Nanotube-Conductive Additive-Space Durable Polymer Nanocomposite Films for Electrostatic Charge Dissipation," *Polymer*, Vol. 45, 2004, pp. 6133–6142.

[5] Peng, H., Alemany, L. B., Margrave, J. L., and Khabashesku, V. N., "Sidewall Carboxylic Acid Functionalization of Single-Walled Carbon Nanotubes," *J. Am. Chem. Soc.*, Vol. 125, No. 49, 2003, pp. 15174–15182.

[6] Chen, J., Rao, A. M., Lyuksyutov, S., Itkis, M. E., Hamon, M. A., Hu, H., Cohn, R. W., Eklund, P. C., Colbert, D. T., Smalley, R. E., and Haddon, R. C., "Dissolution of Full-Length Single-Walled Carbon Nanotubes," *J. Phys. Chem. B*, Vol. 105, No. 13, 2001, pp. 2525–2528.

[7] Koshio, A., Yudasaka, M., Zhang, M., and Iijima, S., "A Simple Way to Chemically React Single-Wall Carbon Nanotubes with Organic Materials Using Ultrasonication," *Nano Lett.*, Vol. 1, No. 7, 2001, pp. 361–363.

[8] Zhang, M., Yudasaka, M., Koshio, A., and Iijima, S., "Thermogravimetric Analysis of Single-Wall Carbon Nanotubes Ultrasonicated in Monochlorobenzene," *Chem. Phys. Lett.*, Vol. 364, 2002, pp. 420–426.

FAST RATE ANALYSIS

Journal of ASTM International, November/December 2005, Vol. 2, No. 10
Paper ID JAI12797
Available online at www.astm.org

Bryan Bilyeu,[1] *Witold Brostow,*[1] *and Kevin P. Menard*[2]

Characterization Of Epoxy Curing Using High Heating Rate DSC

ABSTRACT: Two common techniques to characterize the curing reaction of an epoxy as a function of time and temperature are heat of reaction and the shift in glass transition. These techniques can both be performed by Differential Scanning Calorimetry. However, both techniques share a common limitation, namely that the curing exotherm may occur during the ramp to isothermal temperature or may overlap with the glass transition event. Since the curing reaction and thus the exotherm is a kinetic event, it is rate dependent. As faster rates are used, the scan can be completed before the onset of the reaction. In an isothermal heat of reaction study, the material can be brought to temperature fast enough that the curing reaction does not begin until it is at temperature. Also, since the glass transition can be measured at high scanning rates, it can be measured before the curing reaction begins. At 100°C/min, the two overlapping events are separated, whereas at 200°C/min, the curing does not occur in the temperature range studied. The glass transition temperature can be measured without affecting the degree of cure, or the exotherm can be shifted to separate it from the glass transition for accurate measurement of enthalpy and the glass transition.

KEYWORDS: epoxy, differential scanning calorimetry, exotherm, separation, glass transition

Introduction

Epoxies are versatile and commercially important thermosets with desirable properties [1] due to the nature of the inter-chain bonds and the high degree of crosslinking. These properties also make epoxies excellent matrices for fiber-reinforced composites[2]. A review of epoxy properties and applications is available [3]. The reactive groups of epoxy monomers are three-membered rings of two carbons and an oxygen atom. The reactive epoxide ring is typically opened by an amine or an anhydride curing agent, with amines being more common due to the speed and also due to the ability to modify the main epoxy monomer with a specialty amine [4]. Epoxy curing is a multistep process involving primary and secondary amine hydrogen attack as well as epoxide-epoxide reactions resulting in both polymerization and crosslinking [5].

Since a great advantage of epoxies is that a low viscosity epoxy and curing agent mixture can be cured to a hard insoluble solid in versatile geometries, characterization of the relationship of temperature and time to the degree of cure is very important in processing epoxies. There are a variety of methods to characterize the degree of cure of an epoxy in terms of chemical conversion, thermodynamic properties, and mechanical properties [6]. However, only techniques using Differential Scanning Calorimetry (DSC) are presented in this paper.

Manuscript received 24 May 2004; accepted for publication 20 May 2005; published November 2005. Presented at ASTM Symposium on Techniques in Thermal Analysis: Hyphenated Techniques Thermal Analysis of the Surface and Fast Rate A on 24-25 May 2004 in West Conshohocken, PA.

[1] Laboratory of Advanced Polymers and Optimized Materials, Department of Materials Science and Engineering, University of North Texas, Denton, TX 76203-5310: www.unt.edu/LAPOM/, bwb0001@unt.edu, brostow@unt.edu.

[2] Perkin Elmer Thermal Laboratory, Department of Materials Science, University of North Texas, Denton, TX 76203-5310, kevin.menard@perkinelmer.com.

Epoxy curing involves an increase in both linear molecular weight and crosslink density, both of which result in reduced chain segment mobility. Increasing the linear molecular weight or crosslink density of a polymer chain increases the position of the glass transition temperature, T_g. Many thermosetting polymer systems exhibit a relationship between the T_g and the degree of chemical conversion [7]. Most epoxy-amine systems exhibit a linear relationship, which implies that the change in molecular structure with conversion is independent of the cure temperature [8]. Such a T_g shift often gives better resolution than exothermic enthalpy changes [9]. A review of models and equations relating the degree of cure to physical properties is available [10]. To generate a kinetic equation based on the T_g shift, a graph of T_g versus time graph is generated for a series of isotherms as shown in Fig. 1 and a log time plot (Fig. 2) prepared to generate shift factors. One important note is that each data point in both figures represents one complete scan, i.e., the sample is ramped to a specified isothermal temperature for a predetermined time, then quenched, and a T_g scan performed, and a new sample prepared for the next data point.

The most convenient and generally most accurate method for determining the T_g of polymers is Differential Scanning Calorimetry (DSC) [11]. The T_g is taken as the temperature at the inflection point (peak of derivative curve) [12] of the baseline shift in heat flow or as the temperature at the half height shift in baseline heat flow. In situations where the T_g is distorted or masked by other events, like curing exotherms or enthalpic relaxations, the T_g may be determined by the onset, but results should be noted as onset values to avoid confusion. For consistency in reporting, the ASTM Test Method for Assignment of the Glass Transition (E 1356-98) describes the methods of T_g calculation.

A limitation of DSC is that when measuring an uncured or a partially cured thermoset, a residual exotherm often overlaps the T_g [13]. In addition to earlier methods we developed [14], we have recently developed a method [15] using high heating rate DSC. The transition to the rubber state provides an increase in mobility within the resin, which typically marks the beginning of the curing reaction. As such, the curing exotherm typically begins just after the glass to rubber transition, sometimes even overlapping it. While the shift of the glass transition temperature at high rates due to the time of the molecular motion [16] is small [17], the shift in the exotherm of the curing reaction due to reaction kinetics is large. By increasing the scanning rate, the scan can be completed before the curing reaction and associated exotherm begin. Thus, while the glass transition temperature shift is small, the overlapping exotherm can be pushed to higher temperature, effectively separating the two. In extremely high rates, the scan can be performed without the curing reaction taking place.

A recurring problem in isothermal curing studies is the reaction which takes place during the heating to the curing temperature. Several methods to compensate for this lost heat have been proposed, including dropping the cold sample into a preheated DSC or curve fitting to estimate the lost signal [18]. What is needed is a way to get the sample to the isotherm fast enough to establish temperature control in the furnace before the curing reaction begins. This is a challenge, which requires fast heating and cooling ramps, small sample size, quick thermal response, and equilibration.

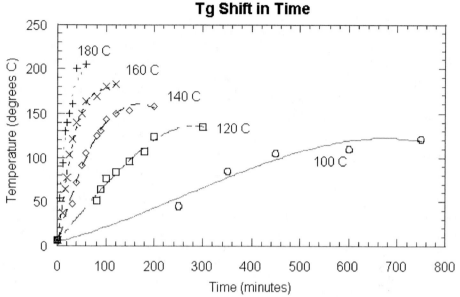

FIG. 1—*Data points showing the change in T_g in time for various curing temperatures [13].*

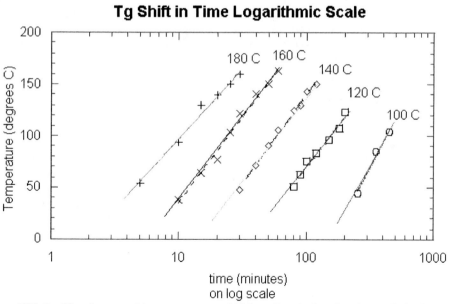

FIG. 2—*The glass transition temperature shift from Fig. 1 plotted on log scale [13].*

Experimental

The commercial DGEBA epoxy System Three cured with the System Three aliphatic amine was used in all measurements. Samples were quickly mixed at room temperature, transferred to aluminum DSC pans, then quickly cooled to –30°C until ready for use. The fully cured samples were held at 100°C for 30 min as per the manufacturer's recommendation. DSC measurements were made on the power compensation Perkin Elmer Diamond DSC operating Pyris software on a Windows platform using liquid nitrogen as the coolant. The instrument was calibrated for temperature and heat of fusion with an indium standard.

Results

The effect of scanning rate on the glass transition temperature and the curing reaction was evaluated. A cured epoxy sample was scanned at various rates between 5 and 200°C/min to verify earlier work [19] on the shift of glass transition temperature with scanning rate. Figure 3 is the series of scans on the same sample at rates of 5, 10, 20, 50, 100, and 200°C/min. The shift in onset and half height glass transition temperatures are not significant within the range of rates. With that established, we turn to the real issue of what can we do with an uncured sample. Figure 4 is a scan of an uncured sample at a standard rate of 40°C/min. While the onset of the glass transition is clear, the end overlaps with the beginning of the curing exotherm, making half height calculation of transition unsure, as well as masking the beginning of the curing exotherm. Increasing the heating rate to 100°C/min in Fig. 5 shifts the onset of the exotherm to a higher temperature. Increasing it further to 200°C/min in Fig. 6 eliminates the curing within the range we are measuring. To determine if the curing reaction was really eliminated from the glass transition measurement at 200°C/min, we performed a series of scans on the same uncured sample to determine if the glass transition temperature changed. In Fig. 7, five consecutive scans of an uncured sample were performed, with no change in the glass transition temperature. In contrast, a series of three consecutive scans at 100°C/min does show a clear increase in the glass transition temperature, indicating some cure is occurring during the scans. Since the reactive epoxy could be scanned for T_g at 200°C/min without affecting the cure state, we developed a new technique for generation of T_g shift in time. Instead of using a new sample for each data point, as was done by traditional methods (shown in Fig. 1), we successfully generated a T_g shift series using only one sample in a single multistep program for each curing temperature. The temperature program and resulting heat flow is shown in Fig. 8. The uncured sample is scanned for T_g and quenched without any curing, then quickly heated to the isothermal temperature and held for a specified time (5 min), then quenched again. A new quick scan is done to measure T_g without affecting the cure state, then quenched. The process is repeated until maximum cure is reached for that temperature, as indicated by a constant or decreasing T_g. Thus, with one sample on one program an entire T_g shift series can be generated for a curing temperature as shown in Fig. 9.

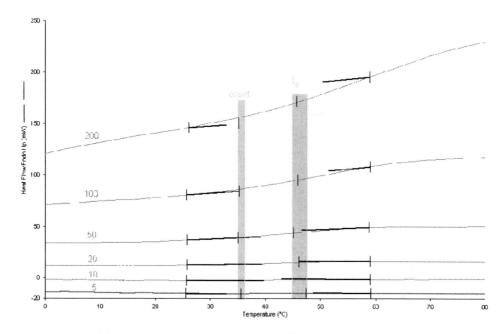

FIG. 3—*Glass transition for a cured epoxy at rates from 5–200°C/min.*

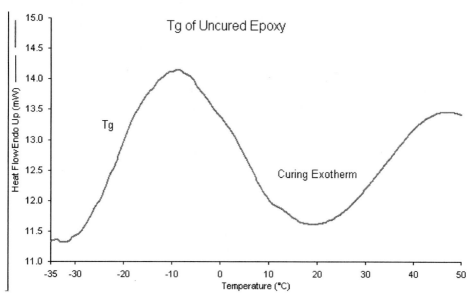

FIG. 4—*DSC scan for glass transition at 40°C/min showing overlapping exotherm.*

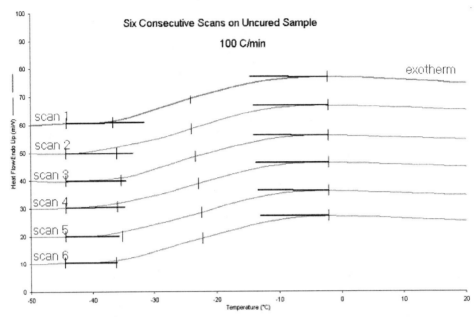

FIG. 5—*Series of six consecutive scans at 100°C/min showing exotherm and advance of T_g.*

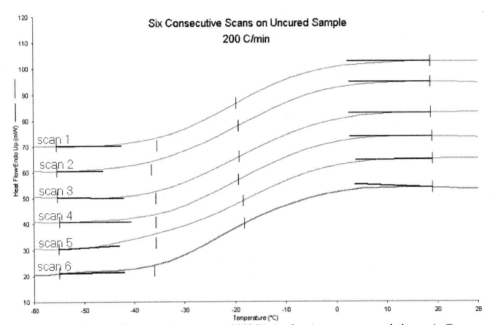

FIG. 6—*Series of consecutive scans at 200°C/min showing no measured change in T_g.*

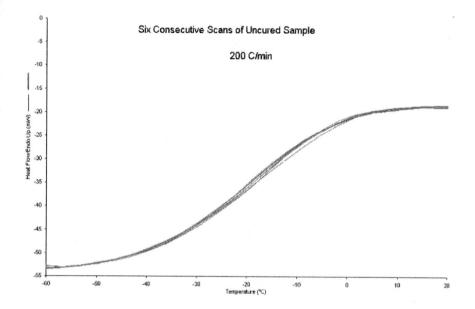

FIG. 7—*Series of scans at 200°C/min, superimposed to show no measured change in T_g.*

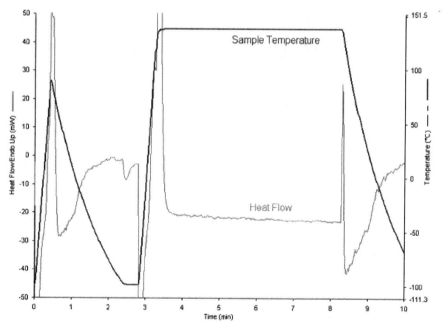

FIG. 8—*Program to generate series for T_g shift data and resulting heat flow.*

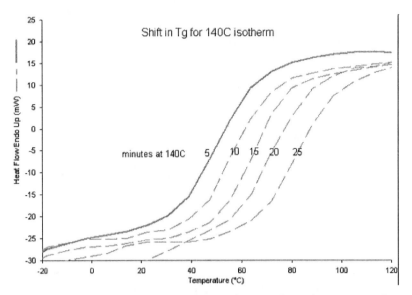

FIG. 9—*Series of T_g values at increasing degrees of cure from one sample.*

Discussion

High scanning rates alleviate many of the limitations inherent to calorimetric study of epoxy curing. The glass transition of an uncured or partially cured epoxy can be determined without affecting curing, the glass transition can be isolated from the curing exotherm, the curing exotherm can be moved away from the glass transition so that the rate dependent kinetics can be calculated or residual cure can be measured, and a sample can be brought to the isothermal cure temperature fast enough to equilibrate before the curing reaction begins. From a practical view, a complete T_g shift in time series can be generated from a single sample on a single multistep program, saving sample material and personnel time that is wasted in traditional methods.

References

[1] Dorman, E. N., *Handbook of Fiberglass and Advanced Plastics Composites*, G. Lubin, Ed., Van Nostrand Reinhold, New York, 1969, p. 46.
[2] Jaffe, M., Menczel, J. D., and Bessey, W. E., *Thermal Characterization of Polymeric Materials*, E. Turi, Ed., Academic Press, San Diego, 1997, p. 1767.
[3] Bilyeu, B., Brostow, W., and Menard, K. P., *Journal of Materials Education*, Vol. 21, 1999, p. 281.
[4] Bilyeu, B., Ph.D. Dissertation, University of North Texas, Dec. 2003.
[5] Stepto, R. F. T., *Cross-Linked Polymers*, R. A. Dickie, S. S. Labana, and R. S. Bauer, Eds., American Chemical Society, Washington DC, 1988, p. 28.
[6] Bilyeu, B., Brostow, W., and Menard, K. P., *Journal of Materials Education*, Vol. 22, 2000, p. 107.

[7] May, C. A., *Applied Polymer Science*, R. Tess and G. Poehlein, Eds., American Chemical Society, Washington DC, 1985, p. 557.

[8] Wisanrakkit, G. and Gillham, J. K., *Polymer Characterization*, C. Craver and T. Provder, Eds., American Chemical Society, Washington DC, 1990, p. 143.

[9] Richardson, M. J., *Pure & Applied Chemistry*, Vol. 64, 1992, p. 1789.

[10] Bilyeu, B., Brostow, W., and Menard, K. P., *Journal Materials Education*, Vol. 23, 2001, p. 189.

[11] Sperling, L. H., *Introduction to Physical Polymer Science*, Wiley, New York, 1986.

[12] Bershtein, V. A. and Egorov, V. M., *Differential Scanning Calorimetry of Polymers*, Ellis Horwood, New York, 1994.

[13] Bilyeu, B., Brostow, W., and Menard, K. P., *Polymer Composites*, Vol. 23, 2002, p. 1111.

[14] Bilyeu, B., Brostow, W., and Menard, K. P., "Materials Characterization by Dynamic and Modulated Thermal Analytical Techniques," *ASTM STP 1402*, A. T. Riga and L. H. Judovits, Eds., ASTM International, West Conshohocken, PA, 2001.

[15] Bilyeu, B., Brostow, W., and Menard, K. P., *Proceedings of the Annual Technical Conference of the Society of Plastics Engineers*, Nashville, Vol. 61, May 2003, p. 1876.

[16] Wunderlich, B., *Thermal Characterization of Polymeric Materials*, E. Turi, Ed., Academic Press, San Diego, 1997.

[17] Chartoff, R. P., *Thermal Characterization of Polymeric Materials*, E. Turi, Ed., Academic Press, San Diego, 1997.

[18] Prime, R. B., *Thermal Characterization of Polymeric Materials*, E. Turi, Ed., Academic Press, San Diego, 1997.

THERMAL ANALYSIS OF THE SURFACE

Journal of ASTM International, November/December 2005, Vol. 2, No. 10
Paper ID JAI12800
Available online at www.astm.org

Carlton G. Slough, Ph.D.,[1] Azzedine Hammiche, Ph.D.,[2] Mike Reading, Ph.D.,[3] and Hubert M. Pollock, Ph.D.[2]

Photo Thermal Micro-Spectroscopy - A New Method for Infared Analysis of Materials

ABSTRACT: Many modern materials are composite structures with complex morphologies that play a large role in determining the material function. The ability to investigate the relationship between structure and property on a microscopic scale can play a crucial role in material development. Micro-Thermal Analysis (μTA)™ is a unique set of analytical techniques for characterizing materials on a micrometer and nanometer scale. Micro-TA combines the imaging power of atomic force microscopy with the ability to analyze physical, mechanical, thermal, and chemical characteristics at a specific point of interest on the surface of a material. At the heart of the technique is a miniaturized thermal probe. In a new technique termed Photo Thermal Micro-Spectroscopy (PTMS), this thermal probe is used to detect temperature fluctuations in samples that have been irradiated by IR radiation. A Fourier Transform Infrared (FT-IR) spectrum can be constructed from this information. This paper describes the PTMS technique and discusses recent applications.

KEYWORDS: Atomic Force Microscopy, Scanning Thermal Microscopy, Photothermal, Infared Spectroscopy, Infared Microscopy, FT-IR, Microspectroscopy

Introduction

Infrared spectroscopy and microscopy are highly utilized in materials analysis for chemical identification. However, there are limitations to the current techniques. Most IR instruments are diffraction limited in resolution to ~10 μm. The possibility to record infared spectra from localized areas smaller than this is highly desirable. Also, in some cases it may be impossible or undesirable to alter the sample using preparation techniques since morphological changes could be incurred. A new technique termed Photo Thermal Micro-Spectroscopy, or PTMS, has demonstrated the ability to collect IR spectra from micrometer-sized areas with little or no sample preparation [1–4]. The ultimate resolution is unknown at this point, but it most certainly will allow collection of spectra at a resolution better than the diffraction limit. It is based on the technique of Micro Thermal Analysis™ (μTA), which uses a heated probe to thermally characterize surfaces on a micrometer level [5].

In μTA, an atomic force microscope (AFM) is modified by replacing the standard silicon probe used for imaging with a miniaturized thermal probe made from Wollaston wire [6]. Wollaston wire is composed of 75 μm diameter Ag wire with a 5 μm Pt/10%Rh core. A tip is formed by bending the wire into a "V" shape and etching the Ag coating away to expose ~50 μm

Manuscript received 11 June 2004; accepted for publication 17 May 2005; published November 2005. Presented at ASTM Symposium on Techniques in Thermal Analysis: Hyphenated Techniques Thermal Analysis of the Surface and Fast Rate A on 24-25 May 2004 in West Conshohocken, PA.
[1] TA Instruments, 109 Lukens Dr., New Castle, DE 19720, USA.
[2] Department of Physics, Lancaster University, Lancaster LA1 4YB, UK.
[3] Department of Chemical Sciences and Pharmacy, University of East Anglia, Norwich NR4 7TJ, UK.

length of the Pt/Rh wire. This exposed wire acts as the sensor. Figure 1 shows an SEM micrograph of the probe. This type of probe can be used for active heating of the surface, by passing a current through it, or passive temperature detection, by measuring the resistance under constant current conditions. It is the latter mode that is used in the PTMS application. In PTMS the thermal probe is used to detect temperature changes in a sample that has been irradiated with IR radiation. In standard µTA the probe, used in the active mode, produces both topography and thermal conductivity images of a surface and performs local thermal characterization on a micrometer scale. Thus, the potential exists to image, thermally characterize, and chemically analyze a surface with one instrument. Ultimately, it should be possible to extend the technique so that IR imaging of sample surfaces can be conducted. In this mode, images would be created where contrast is based upon level of absorption for a particular frequency of IR radiation.

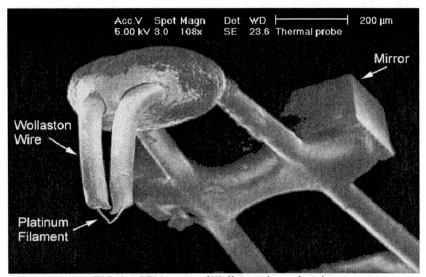

FIG. 1—*SEM image of Wollaston thermal probe.*

The PTMS technique works as follows [3]; Infared radiation from an FTIR spectrometer is used to irradiate a sample surface. Absorption of this radiation by the sample induces temperature fluctuations, which are detected as resistance changes by the thermal probe, acting as a temperature sensor, over time. This signal becomes the interferogram, which is amplified and fed into a spectrometer. A Fourier transform algorithm is applied to deconvolute the spectrum from the interferogram.

The data reported here are based on work performed in the United Kingdom.

Experimental

In the experimental setup, the head from an Explorer AFM is mounted in the chamber of a Bruker Vector 22 FTIR spectrometer. Condensing optics are used to increase the IR flux at the sample position. The tip is positioned at the focal point of the radiation, whose beam is 2 mm in diameter. The sample and microscope are both mounted on a micrometer-based translator capable of X, Y, and Z translation. The inteferogram is fed into the external input of the spectrometer, after passing through a custom built amplifier (Specac, Ltd.), where it is deconvoluted [3].

A typical experiment is conducted by first obtaining a background IR spectrum with the probe far removed from the surface of interest (a few millimeters). A background scan is required since the probe is not only heated by the sample but also by direct absorption of the IR radiation. Typically, only one or two background scans are required for each run. The probe is then lowered onto the surface, and another IR spectrum is recorded. The background is then subtracted to produce the resultant IR spectrum. To pull a useful spectrum out of the noise, coherent averaging by coaddition is performed [4]. With recent improvements in signal to noise ratio, typically a few hundred coadditions are now required. Typical resolution is 4 cm^{-1}.

Results and Discussion

There is, of course, interest in both the lateral resolution of the technique and the depth dependency. Figures 2 and 3 illustrate results on model systems, which address these areas. The left side of Fig. 2 shows two beads of polystyrene of nominal diameter 5 μm imbedded in rock salt (IR transparent) [7]. The thermal probe has been used to image and thus locate the positions of the beads first. After imaging, the probe can be positioned on a bead and an IR spectrum obtained. The ability to image first and then position the probe on an area of interest is itself a unique capability of the instrument. On the right side, the sample spectrum and background are shown along with the final result from their subtraction. Finally, an attenuated total reflection (ATR) IR spectrum for PS is shown. Comparison of the spectra shows good agreement. From this it can be inferred that if a specimen is surrounded by IR transparent material, a good spectrum can be obtained from a very small amount of material.

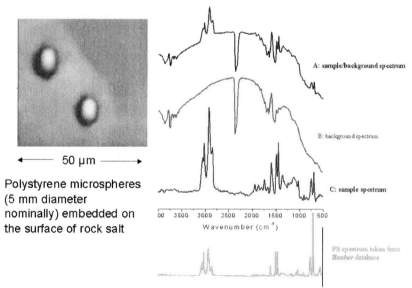

FIG. 2—*Polystyrene microspheres imaged and IR characterized.*

Figure 3 illustrates the depth dependency of the technique [4]. In this experiment, a layer of PS was spin cast onto a glass slide and then coated with thicker and thicker layers of PIB. After each deposit of PIB, a PTMS experiment was performed until the IR signal from the PS was no longer distinguishable. The IR bands used to characterize the PS were the C-H out of plane bending of the aromatic ring for PS (~700–760 cm^{-1}) and the C-H stretch for PIB (~2700–3200 cm^{-1}). The data can be looked at two ways. First, even with the thinnest overlayer of PIB at 2 µm, some features of the PIB are apparent. This provides some information to a typical question in micro thermal analysis of what is the thinnest layer that can be analyzed. Second, focusing on the PS, the data indicate that the PIB overlayer must be greater than 15 µm to lose detection of the PS. Thus, the technique is not restricted to the immediate surface but has subsurface capabilities.

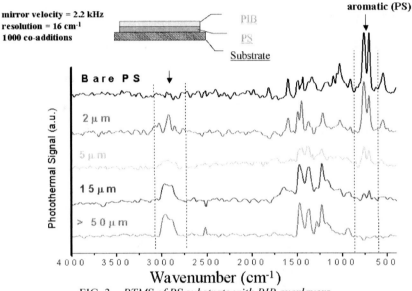

FIG. 3—*PTMS of PS substrate with PIB overlayers.*

Even though the technique is still in its infancy, it has already has shown one clear advantage over conventional FTIR microscopy: the ability to gather IR spectra on difficult samples such as fibers [7]. Many times there is a need to press the sample of interest onto an IR transparent block for placement into the FTIR spectrometer. This can cause unwanted changes in sample morphology, which, in turn will have an impact on the IR spectrum. The PTMS technique avoids this.

Figures 4 and 5 illustrate some of the systems that have been studied to date with the technique. Figure 4 shows the detection of surfactant in a micro-droplet of water and illustrates the sensitivity of the technique. In this case the tip is gently lowered into the droplet and the IR spectrum obtained. Even though water absorbs strongly in the IR region, the IR spectrum clearly shows the surfactant as can be analyzed. Comparison to a conventional ATR spectrum is good.

Figure 5 illustrates the potential ability of the technique to perform a forensic type analysis. In this example a drop of a blend of PS/LDPE (90/10) dissolved in tolulene was absorbed onto a tissue. This can be thought of as representing an "unknown." This was then analyzed with the PTMS technique. A baseline spectrogram of the bare tissue was also collected and subtracted from the spectrogram of the tissue plus liquid. The resultant spectrogram was then compared to the spectra for PS, PE, and tolulene. It is seen that the spectrogram matches the PS spectrogram (characteristic peaks are indicated by arrows). As a control, a chip of the polymer blend was dissolved in tolulene and allowed to dry on a glass slide. The probe then picked up a flake, and PTMS was performed. The "unknown" and the flake spectra compare quite well.

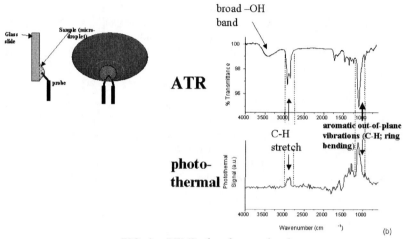

FIG. 4—*PTMS of surfactant droplet.*

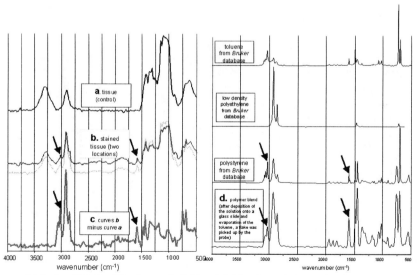

FIG. 5—*PTMS of "unknown" (PS/LDPE blend in tolulene) droplet on tissue.*

Other systems that have been analyzed with the PTMS technique include healthy and malignant prostate cells, layered materials, fibers such as hair, coatings on wires, and DNA. It is clear that the technique is very versatile in its application.

Of special interest is the promise to perform IR mapping. This would involve scanning the probe over the surface and collecting IR spectra at the same time. Images could be created where contrast is based on the intensity at a particular IR wavelength. Standard IR microscopy is limited by the diffraction of the radiation to a resolution of 10 μm. Though other factors will determine the resolution of the PTMS technique, it should, in principle, be able to improve on this limit significantly. Among the factors that will determine the ultimate resolution are the probe size and the sharpness of the temperature distribution. As far as the latter is concerned, some modeling has been done [3]. Results indicated a resolution of at least 100 nm, if just the DC temperature distribution is analyzed. Modulation of the temperature response, which occurs in a typical IR experiment, should improve upon this.

Conclusions

Photo Thermal Micro Spectroscopy is a very promising technique for the imaging and chemical analysis of a wide variety of samples. It has the unique ability to obtain mid-IR spectra from difficult samples such as fibers without use of preparation techniques that could change the sample morphology. Combined with the ability to image the surface prior to obtaining the spectra, the PTMS technique offers some distinct advantages over conventional IR techniques. The next step is to obtain a true IR microscope that operates at a wavelength less than the diffraction limit.

Previous work has included, among other things, the use of synchrotron radiation as an IR source for increased signal to noise [8]. Also, this paper has not covered developments such as nanosampling techniques [9] or the use of high-resolution thermal probes.

References

[1] For further information: http://www.lancs.ac.uk/depts/physics/research/condmatt/micro/micro.html.
[2] "Method and Apparatus for Localized Infrared Spectroscopy Combined with Scanning Probe Microscopy," US patent number 6,260,997.
[3] Hammiche, A., Pollock, H. M., Reading, M., Claybourn, M., Turner, P. H., and Jewkes, K., *Appl. Spec.*, Vol. 53, No. 7, 1999, pp. 810–815.   [CrossRef]
[4] Bozec, L., Hammiche, A., Pollock, H. M., Conroy, M., Chalmers, J. M., Everall, N. J., and Turin, L., *J. Appl. Phys.*, Vol. 90, 2001, pp. 5159–5165.   [CrossRef]
[5] Hammiche, A. and Pollock, H. M., *J. Phys. D.: Appl. Phys.*, Vol. 34, R23-R53, 2001.
[6] Pollock, H. M. and Smith, D. A., *Handbook of Vibrational Spectroscopy*, Chalmers, J. M. and Griffiths, P. M., Eds., John Wiley and Sons Ltd, Vol. 2, 2002, pp. 1472–1492.
[7] Hammiche, A., Bozed, L., German, M. J., Chalmers, M., Everall, N. N., Poulter, G., et al., *Spectroscopy*, Vol. 19, No. 2, 2004, pp. 20–42.
[8] Bozec, L., Hammiche, A., Tobin, M. J., Chalmers, J. M., Everall, N. J., and Pollock, H. M., *Measurement Science and Technology*, Vol. 13, 2002, pp. 1217–1222. [CrossRef]
[9] Reading, M., Grandy, D., Hammiche, A., Bozec, L., and Pollock, H. M., *Vibrational Spectroscopy*, Vol. 901, 2002, pp. 257–260.
Note: The Bruker database was accessed via http://www.brukeroptics.com/.

Journal of ASTM International, April 2006, Vol. 3, No. 4
Paper ID JAI13895
Available online at www.astm.org

Richard E. Lyon,[1] *Richard N. Walters,*[2] *and Stanislav I. Stoliarov*[2]

A Thermal Analysis Method for Measuring Polymer Flammability

ABSTRACT: A thermal analysis method is presented in which controlled heating of polymer samples and complete combustion of the evolved gases are used to separately reproduce the condensed and gas phase processes of flaming combustion in a single laboratory test. Oxygen consumption calorimetry applied to the combustion gas stream gives the heat release rate history of the sample as a function of its temperature. The maximum rate of heat release and the temperature at which it occurs are polymer characteristics related to fire performance and flame resistance.

KEYWORDS: polymer, plastic, fire, flammability, flame, heat release, kinetics, combustion, oxygen consumption, calorimetry

Introduction

A considerable amount of effort has been expended to relate laboratory thermal analyses to flammability [1–16]. The motivation for these studies is the desire for quantitative data to use in materials evaluation and the convenience of testing milligram-sized samples under equilibrium conditions. Most thermal analyses of flammability attempt to relate a single quasi-equilibrium property such as char yield, heat of combustion, or thermal decomposition temperature to fire or flame test performance. Individually, these material properties have found limited success as descriptors of flammability and their interrelationship in the context of fire behavior has remained obscure [17]. The primary obstacles to relating polymer properties to flame and fire test results are the highly coupled nature of the gas and condensed phase processes of flaming combustion (heat and mass transfer), physical changes of the solid during burning (melting, dripping, swelling, char barrier formation), and combustion inhibition in the gas phase due to the presence (halogens) or absence (oxygen) of chemical species in the flame.

In this paper, a test method is described that separately reproduces the condensed phase (pyrolysis) and gas phase (combustion) processes of flaming combustion in a single laboratory test. A simple model of polymer combustion is used to interpret the test results and excellent agreement with the experimental data is observed. A physical basis for thermal analysis of polymer flammability is thus established and a material property is identified that is a good predictor of fire behavior and flame resistance [18–26].

Condensed Phase Model

The burning of a condensed phase material (e.g., a solid polymer) produces volatile fuel species and possibly a solid carbonaceous char or ash under anaerobic conditions [12,13]. The material at the burning surface is heated at a rate that is the product of the surface temperature gradient and the surface recession velocity, and is typically on the order of a few degrees Kelvin per second [18,19]. The process of volatile fuel generation at the burning polymer surface is well described by a single-step, anaerobic thermal decomposition reaction [18,19,27–34]

Manuscript received December 12, 2005; accepted for publication January 19, 2005; published online March 2006. Presented at ASTM Symposium on Techniques in Thermal Analysis: Hyphenated Techniques, Thermal Analysis of the Surface, and Fast Rate Analysis on 24 May 2006 in West Conshohocken, PA; L. H. Judovits and W. P. Pan, Guest Editors.
[1] Fire Safety Branch AAR-440, Federal Aviation Administration, William J. Hughes Technical Center, Atlantic City International Airport, New Jersey 08405.

[2] SRA International, Egg Harbor Township, NJ 08405.

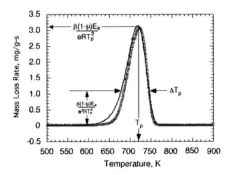

FIG. 1—*Plot of Eq 8 for* $a=5 \times 10^{11}$ s^{-1}, $E_a=190$ kJ/mole *(solid line) compared to experimental data (symbols) for PA66. Peak height (Eq. 7), peak width (Eq 8) and the temperature at maximum mass loss rate* T_p *are indicated by arrows in the figure.*

$$P \rightarrow F(\uparrow) + C \tag{1}$$

where species P, F, and C represent the polymer (P) and its fuel gases (F) and solid thermal decomposition products (C), respectively. It can be shown that if m_P, m_F, and m_C are the masses of P, F, and C, m $=m_P+m_C$ is the sensible mass, $m_0=m_P+m_F+m_C$ is the initial mass, and $\mu=m_C/m_0$ is a constant, then the instantaneous rate of mass loss/fuel generation, is

$$-\frac{dm}{dt} = k_p(m - \mu m_0) \tag{2}$$

In Eq 2,

$$k_p = A \exp\left|-\frac{E_a}{RT}\right| \tag{3}$$

is the pyrolysis (reaction 1) rate constant at temperature T in terms of the frequency factor A, the global activation energy for pyrolysis E_a, and the gas constant R. Although the heating rate varies with depth in a burning polymer, the heating rate at a particular depth (e.g., the surface) is relatively constant during steady burning [18,19]. A constant heating rate, $dT/dt=\beta$, transforms the independent variable from time to temperature in Eq 2, which is then integrated to obtain the fraction of the initial mass (m_0) remaining at temperature T [18,19]

$$\frac{m(T)}{m_0} = \mu + (1 - \mu)e^{-y} \tag{4}$$

The exponent of Eq 4, $y=(ART^2 \exp[-E_a/RT])/(\beta(E_a+2RT))$ is a complex function of temperature. The specific mass loss rate at temperature T for a constant heating rate β is the time derivative of Eq 4,

$$-\frac{1}{m_0}\frac{dm}{dt} = (1-\mu)k_p e^{-y} \tag{5}$$

Figure 1 shows the mass loss rate history of polyhexamethyleneadipamide (PA66) at heating rate β $=10$ K/min measured under anaerobic conditions in a thermogravimetric analyzer/TGA (see Experimental Section). Also plotted in Fig. 1 is the calculated mass loss history at 10 K/min using Eq 5 with A=5 $\times 10^{11}$ s^{-1} and $E_a=190$ kJ/mole determined for PA66 by nonisothermal gravimetric analysis [35]. A reasonably good fit of the experimental data to Eq 5 is obtained for these kinetic parameters that are typical of polymers [36,37].

The maximum specific mass loss rate in a constant heating rate experiment such as that shown in Fig. 1 is found by setting the time derivative of Eq 5 (or the second derivative of Eq 4) equal to zero. The nontrivial ($\mu \neq 1$) result for the value of the rate constant at the peak mass loss rate temperature T_p, is

$$k_p(max) = \frac{\beta E_a}{RT_p^2} \qquad (6)$$

Substituting $k_p(max)$/Eq 6 into Eq 5 gives the maximum specific mass loss rate of a sample that is uniform in temperature and is heated at a constant rate of temperature rise β [18,19],

$$\left.\frac{-1}{m_0}\frac{dm}{dt}\right|_{max} = \frac{\beta(1-\mu)E_a}{e^\gamma RT_p^2} \approx \frac{\beta(1-\mu)E_a}{eRT_p^2} \qquad (7)$$

where, $\gamma = E_a/(E_a + 2RT_p) \approx 1$ for typical $E_a \gg RT_p$. The maximum fractional mass loss rate of PA66 measured experimentally is 3.1 mg/g-s (see Fig. 1) compared to 2.9 mg/g-s calculated from Eq 7 for kinetic parameters, $E_a = 190$ kJ/mole, $\mu = 0$, and $T_p = 720$ K. A characteristic temperature interval for pyrolysis can be defined

$$\Delta T_p \equiv \frac{-\frac{1}{m_0}\int_{T_0}^{T_\alpha}\frac{dm}{dt}dT}{-\frac{1}{m_0}\frac{dm}{dt}\Big|_{max}} = \frac{-\beta\frac{1}{m_0}\int_0^\infty\frac{dm}{dt}dt}{\beta(1-\mu)E_a/eRT_p^2} = \frac{\beta(1-\mu)}{\beta(1-\mu)E_a/eRT_p^2} = \frac{eRT_p^2}{E_a} \qquad (8)$$

Evaluating Eq 5 at $T = T_p \pm \Delta T_p/2 = T_p \pm eRT_p^2/2E_a$ shows that, on average, the mass loss rate falls to 1/e of the maximum value at $\Delta T_p = eRT_p^2/E_a$ for typical [36,37] polymer E_a, T_p, i.e.,

$$\left.\frac{-1}{m_0}\frac{dm}{dt}\right|_{T=T_p\pm\frac{\Delta T_p}{2}} \approx \frac{1}{e}\left(\left.\frac{-1}{m_0}\frac{dm}{dt}\right|_{max}\right) = \frac{\beta(1-\mu)E_a}{e^2RT_p^2} \qquad (9)$$

The pyrolysis interval $\Delta T_p = eRT_p^2/E_a$ is also plotted in Fig. 1.

The heating rate dependence of T_p is obtained by setting Eqs 3 and 6 equal at $T = T_p$ during a constant heating rate experiment

$$\frac{\beta E_a}{ART_p^2}\exp\left|\frac{E_a}{RT_p}\right| = 1 \qquad (10)$$

The frequency factor A can be eliminated from Eq 10 by defining a reference heating rate β_0 for which $T_p(\beta_0) = T_{p,0}$

$$\frac{\beta E_a}{ART_p^2}\exp\left|\frac{E_a}{RT_p}\right| = \frac{\beta_0 E_a}{ART_{p,0}^2}\exp\left|\frac{E_a}{RT_{p,0}}\right| = 1 \qquad (11)$$

The temperature at maximum specific mass loss rate $T_p(\beta) = T_p$ at heating rate β is obtained from Eq 11 as

$$\frac{1}{T_p} = \frac{1}{T_{p,0}} + \frac{R}{E_a}\ln\left[\frac{\beta_0}{\beta}\right] - \frac{R}{E_a}\ln\left[\frac{T_{p,0}}{T_p}\right]^2 \qquad (12)$$

Changes in T_p are small relative to changes in β for typical [36,37] polymer activation energies $E_a \approx 200 \pm 50$ kJ/mole and decomposition temperatures [36–38] $T_p \approx 700 \pm 50$ K, so the last term on the right-hand side of Eq 12 can be neglected, and

$$\frac{1}{T_p} \approx \frac{1}{T_{p,0}} + \frac{R}{E_a}\ln\left[\frac{\beta_0}{\beta}\right] \qquad (13)$$

Substituting Eq 13 into Eq 7 gives an explicit result for the maximum mass loss rate in a constant heating rate experiment in terms of the kinetic parameters

$$\left.\frac{-1}{m_0}\frac{dm}{dt}\right|_{max} = \frac{\beta(1-\mu)E_a}{eRT_{p,0}^2}\left\{1 + \frac{2RT_{p,0}}{E_a}\ln\left[\frac{\beta_0}{\beta}\right] + \left(\frac{RT_{p,0}}{E_a}\ln\left[\frac{\beta_0}{\beta}\right]\right)^2\right\} \qquad (14)$$

Defining, $x \equiv \beta_0/\beta$ and $a \equiv 2RT_{p,0}/E_a$ and substituting these into Eq 14 shows that the bracketed quantity resembles the series expansion for an exponential

$$x^a = 1 + a\ln[x] + (a\ln[x])^2/2\,! + (a\ln[x])^3/3\,! + \cdots$$

For common polymers [36–38] under typical [39,40] experimental conditions of thermogravimetry, a $= 2RT_{p,0}/E_a \approx (2)(8.314\ \text{J/mole-K})(700\pm50\ \text{K})/(200\pm50\ \text{kJ/mole}) = 0.06\pm0.01$. Hence, $a \ll 1$ and since $\ln[x] = \ln[\beta_0/\beta]$ is of unit order for the decade range of heating rates encountered in thermogravimetric analyses (and fires), both x^a and the bracketed quantity in Eq 14 can be approximated with a sufficient accuracy by the first two terms of the series, which simplifies the heating rate dependence of the maximum specific mass loss rate

$$\left. \frac{-1}{m_0}\frac{dm}{dt}\right|_{max} = \frac{\beta(1-\mu)E_a}{eRT_p^2} = \frac{\beta_0(1-\mu)E_a}{eRT_{p,0}^2}\left[\frac{\beta}{\beta_0}\right]^{1-a} = \left.\frac{-1}{m_0}\frac{dm}{dt}\right|_{max,0}\left[\frac{\beta}{\beta_0}\right]^{1-a} \qquad (15)$$

Multiplying Eq 15 by the heat of combustion of the pyrolysis products $h_{c,v}^0$ gives the maximum specific heat release rate (W/kg) of a polymer sample whose temperature is uniform and increases at a constant rate during which all of the pyrolysis gases are completely and instantaneously combusted

$$Q_{max}(\beta) = \left.\frac{-h_{c,v}^0}{m_0}\frac{dm}{dt}\right|_{max} = \left.\frac{-h_{c,v}^0}{m_0}\frac{dm}{dt}\right|_{max,0}\left[\frac{\beta}{\beta_0}\right]^{1-a} = Q_{max}(\beta_0)\left[\frac{\beta}{\beta_0}\right]^{1-a} \qquad (16)$$

Dividing the maximum specific heat release rate Q_{max} (Eq 16) by β yields

$$\frac{Q_{max}(\beta)}{\beta} = \frac{Q_{max}(\beta_0)}{\beta_0}\left(\frac{\beta_0}{\beta}\right)^a = \eta_c\left(\frac{\beta_0}{\beta}\right)^a \qquad (17a)$$

or,

$$\eta_c = \frac{h_{c,v}^0(1-\mu)E_a}{eRT_{p,0}^2} = \frac{Q_{max}(\beta)}{\beta}\left[\frac{\beta}{\beta_0}\right]^a \qquad (17b)$$

Substituting Eq 8 into Eq 17b,

$$\eta_c = \frac{h_c^0}{\Delta T_{p,0}} = \frac{Q_{max}(\beta_0)}{\beta_0} \qquad (18)$$

Equation 18 shows that η_c is defined as the average amount of heat released by combustion of the pyrolysis gases per degree of temperature rise over the pyrolysis interval at a specific heating rate $\beta = \beta_0$. However, η_c can be calculated from the data obtained at different heating rates using Eq 17b, and it has the units and significance of a heat [release] capacity [18–23]. For a polymer that decomposes by a first order (single step) process, the heat release capacity η_c is seen to be a particular function of thermal stability (E_a, T_p) and combustion ($\mu, h_{c,v}^0$) properties, each of which is known to be separately calculable from additive molar group contributions [37,38,41]. Consequently, η_c should be (and is) calculable from additive molar group contributions [23].

Gas Phase Model

The reaction of volatile fuel F (e.g., Eq 1) with oxygen typically yields complete (CO_2, H_2O, HX) and incomplete (CO, HC) combustion products, where X is a halogen, HX is a halogen acid, and HC is a solid or gaseous hydrocarbon

$$F + gO_2 \rightarrow aCO_2 + bCO + cH_2O + hHX + eHC \qquad (19)$$

Rarely is the stoichiometric oxygen/fuel ratio known in advance and combustion is never 100 % complete during the burning of polymers because of kinetic and diffusion limitations in the gas phase. In Eq 19, $g = a + b/2 + c/2$, and the rate of fuel consumption by oxidation (assuming the second order kinetics) is:

$$-\frac{d[F]}{dt} = k_c[F][O_2] \qquad (20)$$

where [F] and [O_2] are the molar concentrations of fuel and oxygen, respectively, in the gas phase and k_c is the global rate constant for combustion. For combustion in a large excess of oxygen where [O_2] $\approx [O_2]_0$ is approximately constant, Eq 20 becomes

$$-\frac{d[F]}{dt} = \{k_c[O_2]_0\}[F] = k_{app}[F] \tag{21}$$

$k_{app}=[O_2]_0k_c$ is an apparent rate constant for fuel combustion. Equation 21 is solved immediately for the isothermal fuel concentration at time t

$$\frac{[F]}{[F]_0} = 1 - \chi = e^{-k_{app}t} \tag{22}$$

where, $\chi=\chi(t,T)$ is the extent of reaction expressed as the change in fuel concentration $\Delta[F]$ at elapsed time t, temperature T, divided by the change in fuel concentration for complete reaction $\Delta[F]_0$. The relationship between χ and the oxygen consumed by combustion follows directly from Eq 19,

$$\chi = \chi(t,T) = \frac{\Delta[F]}{\Delta[F]_0} = \frac{g\Delta[O_2]}{g\Delta[O_2]_0} = \frac{\Delta[O_2](t,T)}{\Delta[O_2](max)} \tag{23}$$

If oxygen is present in large excess and there is sufficient time and temperature for complete combustion then $\chi=1$, $[F]=0$ and fuel F is quantitatively converted to CO_2, H_2O and possibly HX. For complete combustion the amount of oxygen consumed is uniquely related to the fuel composition, $F=C_cH_hO_mN_nX_x$,

$$C_cH_hO_mN_nX_x, + \left(c + \frac{h-x-2m}{4}\right)O_2 \rightarrow cCO_2 + \frac{h-x}{2}H_2O + \frac{n}{2}N_2 + xHX \tag{24}$$

The stoichiometric oxygen/fuel mass ratio r_0 is readily calculated from Eq 24 for fuels of known composition and is in the range $r_0=2.0\pm1.5$ for the majority of organic compounds [42]. Thornton [43] was the first to notice that the heat of combustion of organic gases and liquids h_c^0 (J/g-fuel) divided by the stoichiometric mass ratio was essentially constant and independent of the type of fuel

$$C = h_c^0/r_0 = 13.1 \pm 0.7 \text{ MJ/kg-}O_2 \tag{25}$$

This observation was extended to solids by Huggett [44] and became the basis for oxygen consumption calorimetry [45,46], whereby measurement of the mass of oxygen consumed from the combustion atmosphere is used to deduce the amount of heat released during the burning of materials and products [47,48]. Equation 25 is valid only for complete combustion, i.e., Eq 24.

Experimental

Materials

Polymers tested in our laboratory were unfilled, natural, or virgin resins obtained from Aldrich Chemical Company, Scientific Polymer Products, original manufacturers, and plastics suppliers. Methane, oxygen, and nitrogen gases used for calibration and testing were dry, ultra high purity (>99.5 %) grades obtained from Matheson Gas Products.

Methods

Thermogravimetry—Thermogravimetric analyses were performed at various heating rates but typically at $\beta=10$ K/min under nitrogen flow of 80 cm^3/min in commercial instruments (STA-851e, Mettler Toledo or TGA-7, Perkin Elmer) using a standard method [40]. Sample mass was between 1 and 5 mg in all cases.

Heat of Combustion—Net heats of complete combustion of solid polymers [41,42,49] were determined on 1 g samples tested in triplicate using high-pressure oxygen bomb calorimetry according to a standard method [50]. The net heat of combustion was determined from the gross calorific value by subtracting the heat of vaporization of water for these polymers of known composition.

Thermal Oxidation of Fuel Gases—Thermal oxidation kinetics of fuel gases were studied to determine the time-temperature requirements for complete combustion of polymer pyrolysis products under laboratory conditions. In these experiments methane and polymer (PMMA, PP) pyrolyzates were mixed with twice the amount of oxygen required for complete oxidation to carbon dioxide and water, e.g., Eq 19 with $b=h=e=0$. The apparatus used for the oxidation kinetic experiments has been described previously [45,46] and consists of a pyrolysis probe (Pyroprobe 2000, CDS Analytical) in a heated manifold attached to a 5 m long Inconel combustion tube having an inner diameter of 4.5 mm which is coiled to fit inside of a ceramic furnace. The oxygen/fuel ratio was held constant at $[O_2]_0/(g[F]) \geqslant 2$ so that oxygen was always present in excess while the residence time of the gases and the temperature of the combustor were independently varied in stepwise increments of 50 cm^3/min between flow rates of 50–200 cm^3/min and in 10°C increments between combustor temperatures of 500–1000°C. The oxidized gas stream was analyzed for residual oxygen to compute the extent of reaction χ as per Eq 23 for a particular time and temperature in the combustor. Fuel gases tested were methane (4 % by volume in air), the volatile pyrolysis products of polymethylmethacrylate/PMMA (which depolymerizes to methylmethacrylate monomer), and the pyrolysis products of polypropylene (which thermally degrades by random and beta scission to alkanes and alkenes).

Equation 22 was used to model the isothermal oxidation kinetics as a single step process with apparent rate constant

$$k_{app} = [O_2]_0 A_c \exp\left[-\frac{E_c}{RT} \right] \tag{26}$$

where A_c (m^3/(mol-s)) is the pre-exponential factor and E_c (kJ/mol) is the activation energy for the fuel-oxygen reaction, and $[O_2]_0 = 8.6$ mol/m^3 is the oxygen concentration used in the experiments. According to Eqs 22–24, the slope of a plot of $-\ln[1-\chi]$ versus time t at constant temperature T is the apparent rate constant $k_{app}(T)$. The oxidation kinetic parameters E_c and A_c can then be determined from the slope and intercept, respectively, of a plot of $\ln[k_{app}(T)]$ versus $1/T$ as per Eq 26 written in logarithmic form,

$$\ln\{k_{app}(T)/[O_2]_0\} = \ln A_c - (E_c/R)(1/T) \tag{27}$$

Pyrolysis Combustion Flow Calorimetry (PCFC)—Thermal analysis of polymer flammability was conducted using pyrolysis combustion flow calorimetry/PCFC [51–55]. The PCFC methodology, shown schematically in Fig. 2, uses oxygen consumption calorimetry [45,46] to measure the rate and amount of heat produced by complete combustion of the fuel gases generated during controlled pyrolysis of milligram-sized samples. The method is implemented as a stand-alone device as shown in Fig. 3 or as an evolved gas accessory attached to a thermogravimetric analyzer (TGA). In the stand-alone apparatus, 1–5 milligram samples are heated to 800°C at a heating rate of 1°C/s (typically) in a stream of nitrogen flowing at 80 cm^3/min. The volatile thermal degradation products are swept from the pyrolyzer by the nitrogen purge gas and mixed with 20 cm^3/min of pure oxygen prior to entering the combustor held at 900°C (see Results/Thermal Oxidation Kinetics of Combustible Gases). After exiting the combustor the gas stream passes over anhydrous calcium sulfate (Drierite) to remove moisture and acid gases prior to

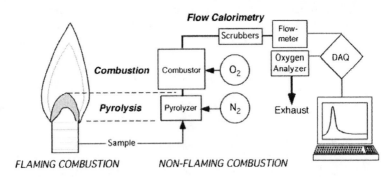

FIG. 2—*Flow diagram for pyrolysis-combustion flow calorimetry.*

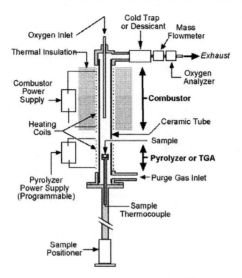

FIG. 3——*Schematic drawing of PCFC.*

passing through a mass flow meter and oxygen analyzer to calculate the heat release rate by oxygen consumption.

Experiments were also conducted in which the combustor was attached to the furnace of a thermogravimetric analyzer (STA-851e, Mettler-Toledo) to thermally oxidize the evolved pyrolysis gases. Three to five samples of each polymer were tested. The heat release rate data were synchronized with the sample temperature by subtracting the transit time of the gases from the pyrolyzer (PCFC) or TGA furnace (STA) to the oxygen analyzer.

Experiments were also conducted in which the purge gas was air rather than nitrogen to effect oxidative pyrolysis. In these experiments both the sample gases and the solid pyrolysis residue/char are completely oxidized and the net heat of combustion of the entire sample is measured by oxygen consumption.

Experiments were also conducted in which the purge gas was methane (8.3 cm^3/min) and nitrogen (75 cm^3/min) and the oxygen flow rate was 16.7 cm^3/min so that $[O_2]/[CH_4]=1$. The combustor temperature was slowly cycled between 25 and 950°C so that the temperature of the $CH_4/O_2/N_2$ gas mixture did not change significantly during the 10 s residence time in the combustor. The oxidized gas stream was analyzed for residual oxygen to compute the extent of reaction as a function of combustor temperature for a residence time of 10 s.

The specific heat release rate is calculated from Eqs 16 and 25 for an initial sample mass m_0, in terms of the instantaneous change in the mass fraction of oxygen in the dried combustion gas stream Δm_{O_2}, the dry gas stream density ρ (kg/m^3), and the volumetric flow rate F (m^3/s),

$$Q(t) = \frac{-h_{c,v}^0}{m_0}\frac{dm}{dt} = \frac{C}{m_0}\frac{dr_0 m}{dt} = \frac{\rho CF}{m_0}\Delta m_{O_2}(t) \qquad (28)$$

where $C=13.1\pm0.6$ MJ/kg-O_2 is essentially the heat of combustion of oxygen with typical organic fuels. The heat of combustion of the fuel gases per unit initial mass of sample h_c^0 (J/g)$=(1-\mu)h_{c,v}^0$ is obtained directly by time-integration of Q(t) over the entire test. The char fraction μ, is obtained by weighing the sample before and after the test. The heat release capacity η_c (J/g-K) is obtained by dividing the maximum value of the specific heat release rate Q_{max} by the heating rate in the test β.

Results

Thermogravimetry (Anaerobic Conditions)

Figure 4 shows experimental data [27] for the maximum specific mass loss rate of polymethylmethacrylate/PMMA ($E_a=160$ kJ/mol, $\mu=0$), polyethylene/PE ($E_a=264$ kJ/mol, $\mu=0$), and

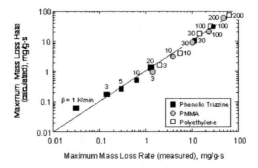

FIG. 4—*Calculated (Eq 7) versus measured peak mass loss rates in TGA for PMMA, polyethylene, and phenolic triazine at heating rates,* $\beta = 1, 3, 5, 10, 20, 30, 100,$ *and* 200 *K/min.*

phenolic triazine/PT ($E_a = 178$ kJ/mol, $\mu = 0.7$) versus heating rate in nitrogen compared to values calculated using Eq 7 with $T_p(\beta)$ measured during the test. Close proximity of the calculated and measured peak mass loss rates to the equivalence line on the log-log plot indicates that Eq 7 is valid over several decades of heating rate in the vicinity of fire heating rates ($1-10$ K/s) if the temperature at maximum mass loss rate $T_p(\beta)$ is used in the calculation with the activation energy and char yield for the polymer determined in separate experiments.

Figure 5 shows T_p versus heating rate data [27] for PE, PT, and PMMA as filled and open circles. Solid lines through the experimental data in Fig. 5 were calculated using Eq 13 with $\beta_0 = 10$ K/min and activation energies obtained by nonisothermal methods [27,35], $E_a = 264, 178,$ and 160 kJ/mol for PE, PT, and PMMA, respectively. The corresponding char yields and peak decomposition temperatures are, $\mu = 0, 0.7, 0,$ and $T_{p,0} = 757, 735,$ and 653 K for PE, PT, and PMMA, respectively.

Figure 6 is a plot of the maximum specific mass loss rate versus heating rate for PE, PT, and PMMA obtained by TGA under nitrogen purge. Symbols are experimental data. Solid lines were calculated using Eq 15 for a reference heating rate $\beta_0 = 10$ K/min.

Thermal Oxidation Kinetics of Combustible Gases

Experimental results obtained in our laboratory for $A = A_c[O_2]_0$ and E_c from thermal oxidation studies of methane (methane 1) and the pyrolyzates of polymethylmethacrylate (PMMA) and polypropylene (PP), are listed in Table 1. Also listed in Table 1 are values of A and E_c obtained from the literature for methane (methane 2) [56] as well as the pyrolysis products of some common hydrocarbon polymers [57].

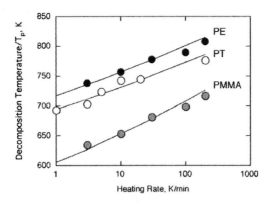

FIG. 5—*Decomposition temperature* $T_p(\beta)$ *versus heating rate* β *for PE, PT and PMMA. Circles are experimental data. Solid lines are obtained from Eq 13 with reported* E_a *and* $\beta_0 = 10$ *K/min.*

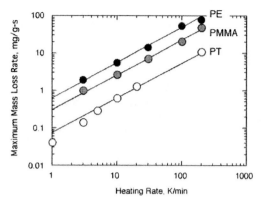

FIG. 6—*Maximum specific mass loss rate versus heating rate for PE, PMMA, and PT polymers. Circles are experimental data. Solid lines are obtained from Eq 15 with $\beta_0 = 10$ K/min.*

From Eqs 22 and 25 and the data in Table 1 the minimum residence time in the combustor at temperature T for any degree of oxidation can be calculated. If the oxidation reaction of the fuel gases in the presence of excess oxygen is required to be 99.5 % complete by the time the gas stream exits the combustor, then the minimum residence time τ_r in the combustor at temperature T is

$$\tau_r = \frac{-\ln(1 - 0.995)}{A \exp[-E_c/RT]} = \frac{5.3}{A \exp[-E_c/RT]} \tag{29}$$

Equation 29 is plotted in Fig. 7 as reaction time τ_r versus temperature for the materials and kinetic parameters in Table 1. Figure 7 shows that for all of the fuels examined thermal oxidation is essentially complete in one second at 1000°C, or in 10 s at 900°C, without the use of a catalyst. These results are significantly different from the 50 seconds at 1000° Celsius claimed by Babrauskas et al. [58] to be necessary for complete thermal oxidation of fire gases containing soot particles using a platinum catalyst.

The results of combustor temperature cycling experiments for the stoichiometric mixture of methane and oxygen in nitrogen are shown in Fig. 8 as the oxygen concentration of the combustion stream exiting the combustor versus the combustor temperature over the range 500–900°C. It is apparent that the oxygen concentration goes to zero, i.e., all of the oxygen (and methane) is consumed during the 10 s residence time in the combustor at temperatures between 775 and 800°C. This result is in general agreement with the Methane 2 data in Fig. 7, although the Methane 1 data indicate that a combustor temperature of 900°C is required for complete oxidation of methane in 10 s. The absence of any residual oxygen in the stoichiometric reaction with methane shows that oxygen is not rate limiting under the conditions of these experiments. The hysteresis in the $[O_2]$ versus time data is due to thermal lag of the temperature measurement.

Table 2 lists the heats of combustion of the pyrolysis products (monomers and oligomers) of non-charring polymers measured in the PCFC for a residence time of 10 s at 900°C in the combustor. Also listed in Table 2 are heats of complete combustion of the same polymers obtained by adiabatic, high-

TABLE 1—*Oxidation kinetic parameters determined experimentally and obtained from the literature [53,54] for methane gas and some polymer pyrolysis products.*

Polymer	E_c (kJ/mole)	A (s^{-1})	Temperature Range (K)	Ref.
Methane gas (methane 1)	241	10^{12}	1020–970	...
Methane gas (methane 2)	230	10^{10}	1000–2000	54
Polymethylmethacrylate (PMMA 1)	62	10^{4}	725–973	55
Polymethylmethacrylate (PMMA 2)	130	10^{7}	773–898	...
Polypropylene	94	10^{5}	607-656	...
Polybutadiene (PB)	91	10^{5}	800-945	55
Polyisoprene (PIS)	75	10^{4}	825-975	55
Ethylene-propylene rubber (EPR)	133	10^{8}	800-975	55
PC/ABS blend	188	10^{10}	800-975	55

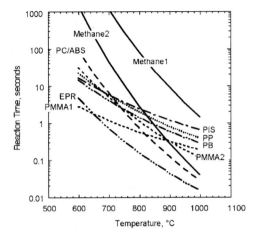

FIG. 7—*Reaction time versus temperature for methane gas and PMMA, PIS, PB, PP, EPR, and PC/ABS polymers calculated from oxidation kinetic parameters.*

pressure, oxygen bomb calorimetry [41,49]. The excellent agreement between PCFC and oxygen bomb calorimetry confirms complete (100 %) combustion of typical polymer pyrolysis products in 10 s at 900°C in excess oxygen.

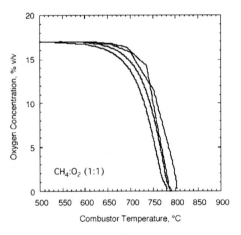

FIG. 8—*Oxygen concentration of a stoichiometric $CH_4:O_2$ mixture in nitrogen exiting the combustor at the indicated temperature.*

TABLE 2—*Net heats of combustion of noncharring polymer pyrolyzates by PCFC compared to oxygen bomb calorimeter values.*

Polymer	ASTM D 2015 (kJ/g)	PCFC (kJ/g)	Percent Relative Deviation
Polyethylene	43.3	43.5±0.1	0.5
Polystyrene	39.8	39.4±0.5	−1.0
Polymethylmethacrylate	24.9	25.0±0.1	0.4
Polyoxymethylene	15.9	16.0±0.1	0.6

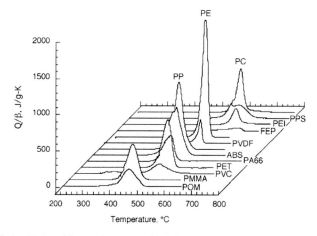

FIG. 9—*Reduced heat release rate histories of common polymers in PCFC.*

Pyrolysis-combustion Flow Calorimetry

Figure 9 shows experimental data for the normalized heat release rate Q/β versus temperature for poly-oxymethylene (POM), PMMA, polyvinylchloride (PVC), polyethyleneterephthalate (PET), polyamide 66 (PA66), ABS, polypropylene (PP), polyvinylidene fluoride (PVDF), polyethylene (PE), fluorinated ethylene-propylene (FEP), polyetherimide (PEI), polyphenylenesulfide (PPS), and polycarbonate (PC) measured by PCFC at a heating rate of 1 K/s using a combustor residence time of 10 s at 900°C. The data in Fig. 9 are sorted from front to back by the temperature at maximum decomposition rate T_p and shows that η_c (Q_{max}/β) varies widely in magnitude and temperature for common polymers.

Figure 10 is a plot of the maximum specific heat release rate Q_{max} versus heating rate for milligram samples of PE, polystyrene/PS, PA66, PMMA, polybutyleneterephthalate/PBT, PET, polyphenleneoxide/PPO, PC, POM and PT. Symbols are experimental data and solid lines are calculated from Eq 16 for typical value $a = 0.06$ and $\beta_0 = 1$ K/s. Slight negative deviation of Q_{max} versus β from proportionality is expected (Eq 16) and observed. However, for the range of heating rates encountered in thermal analyses ($\beta = 0.1 - 1$ K/s) and fires ($\beta_s = 1 - 10$ K/s) the maximum deviation of Q_{max} from proportionality to β is less than 14 %, i.e., $[\beta_0/\beta]^a = [\beta_s/\beta_0]^a \approx [10/1]^{\pm 0.06} = 1.00 \pm 0.14$ for a reference heating rate $\beta_0 = 1$ K/s. The weak dependence of Q_{max}/β on β is illustrated in Fig 11, which shows these data for PE, high impact polystyrene (HIPS), PMMA, POM, and FEP. Symbols are experimental data and solid lines are calculated from Eq 17 for $\beta_0 = 1$ K/s.

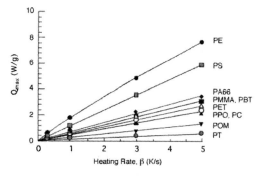

FIG. 10—*Maximum specific heat release rate Q_{max} versus heating rate β in PCFC for 1 mg samples of PE, PS, PA66, PMMA, PBT, PPO, PC, POM, and PT. Symbols are experimental data. Solid lines are obtained from Eq 16 with $a = 0.06$, $\beta_0 = 1$ K/s.*

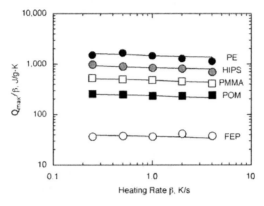

FIG. 11—Q_{max}/β versus β for PE, HIPS, PMMA, POM and FEP. Symbols are experimental data. Solid lines are obtained from Eq 17a.

The repeatability (intra-laboratory variation) of measurements made in our laboratory in the apparatus of Fig. 3 is indicated by the data in Table 3,which lists mean values and one standard deviation for triplicate determinations of heat release capacity η_c, total heat released by combustion of volatile fuel h_c^0, char yield μ, and heat release temperature T_p of the 14 commercial polymers whose heat release rate histories are shown in Fig. 9. Repeatability error estimated from the average coefficient of variation for the data in Table 3 is less than 5 % (i.e., the average relative deviation from the mean is less than 5 %). The reproducibility (inter-laboratory) error for these same polymers obtained using the apparatus of Fig. 3 is about 10 %, as demonstrated graphically in Fig. 12, which is a plot of individual η_c from each of three different laboratories versus the average η_c for the three labs.

To verify the PCFC method, the heat release capacities of 15 polymers measured by PCFC were compared to those measured for the same samples using a thermogravimetric analyzer (TGA) coupled to a gas chromatograph (GC) and mass spectrometer (MS) to determine the fuel species [59–61]. In the TGA-GC/MS method of determining η_c the thermal decomposition products at maximum mass loss rate are sampled, separated, and analyzed by GC/MS and the resulting data used to compute the heat of complete combustion of the fuel gases $h_{c,v}^0$ from their known or calculated heats of combustion and relative abundance (mass fraction). The heat of combustion so determined is multiplied by the maximum value of the fractional mass loss rate measured in the TGA at a constant heating rate (e.g., 10 K/min) to obtain the heat release capacity. The heat release capacities normalized to $\beta = 1$ K/s measured by PCFC and TGA-

TABLE 3—Flammability parameters η_c, h_c^0, μ and T_p (triplicate determinations).

Polymer	η_c (J/g-K)	h_c^0 (kJ/g)	μ (%)	T_p (°K)
HDPE	1486±20	43.5±0.1	0.1±0.1	504±1
PP	1130±24	43.2±0.2	0.0±0.0	483±1
HIPS	859±4	37.8±0.1	2.5±0.2	452±1
PA66	623±34	29.4±0.1	1.0±0.1	475±2
ABS	581±14	37.0±0.2	6.2±0.3	454±1
PC 2	539±26	20.4±0.2	22.5±0.8	547±2
PC 1	484±13	20.4±0.1	23.2±0.2	545±3
PMMA	475±6	24.9±0.1	0.0±0.0	393±2
PET	357±16	16.8±0.7	12.6±1.5	459±3
PPS	248±27	15.7±0.1	44.0±0.6	535±1
POM	235±19	14.3±0.0	0.0±0.0	398±6
PEI	201±7	9.3±0.2	51.3±0.3	565±1
PVC	129±3	10.8±0.2	18.8±0.1	467±4
FEP	57±1	4.1±0.0	0.0±0.0	589±1

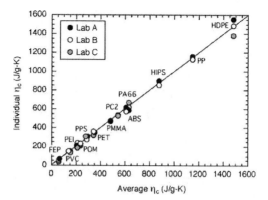

FIG. 12—*Comparison of individual and average heat release capacities from three different laboratories for the 14 polymers in Table 3.*

GC/MS on samples of the same polymer are plotted on the ordinate and abscissa, respectively, in Fig. 13. The proximity of η_c obtained by PCFC and TGA-GC/MS to the equivalence line in Figure 13 corresponds to an average relative deviation from the mean of 13 %.

Figure 14 is a plot of heat release capacity η_c versus $h_c^0/\Delta T_p$ for polymers and commercial plastics that gave a single heat release rate peak centered at Q_{max} with ΔT_p the pyrolysis temperature interval at $Q_{max/e}$. Excellent correlation is observed between heat release capacities obtained by the peak height and peak area methods as per Eq 18.

Experimental data for thermal oxidation of the pyrolysis gases evolved from polycarbonate in the TGA at a heating rate $\beta = 20$ K/min is shown in Fig. 15. Residual mass plotted on the right ordinate shows that thermal decomposition begins at about 450°C and that 24 % of the original mass is left as char at the end of the experiment (700°C). The heat of combustion of the thermal decomposition products is obtained by dividing the specific heat release rate Q(t) by the specific mass loss rate $(m_0^{-1}dm/dt)$ at each temperature T during the test. Figure 15 shows that $h_{c,v}^0$ so obtained ranges from 20–25 kJ/g for the primary decomposition step at 535 ± 25°C that generates monomer fragments (phenol, bisphenol, diphenylcarbonate) and a solid primary char [5,15,62]. The primary char decomposes in a second step to a carbon-rich solid over a broad temperature range with the evolution of methane gas [62], which is consistent with the data in Fig. 15 showing that the heat of combustion of the gases evolved between 550–700°C is on the order of methane ($h_{c,v}^0 = 50$ kJ/g).

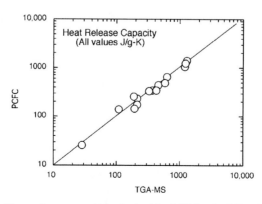

FIG. 13—*Comparison of heat release capacities obtained by PCFC at* $\beta = 260$ *K/min and TGA-GC/MS at* $\beta = 10$ *K/min.*

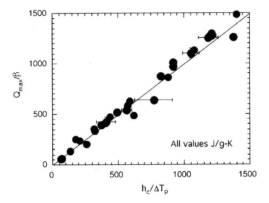

FIG. 14—*Peak height versus peak area method of calculating the heat release capacity.*

Oxidative Pyrolysis-Combustion Flow Calorimetry

Figure 16 shows experimental data from oxidative pyrolysis-combustion flow calorimetry (oPCFC) in the apparatus of Fig. 3 for a 1 mg sample of polycarbonate at $\beta = 5$ K/s. Oxidation of the sample gases in the combustor and the delayed oxidation of the solid char in the pyrolyzer during an air purge are shown as separate processes. The area under the Q(t) versus time curve is the net heat of complete combustion of polycarbonate, $h_c^0 = 29.1$ kJ/g in this case. Table 4 compares data for the net heat of combustion of several polymers obtained by oxygen bomb calorimetry [42,43] and oxidative pyrolysis-combustion flow calorimetry. The error of the oPCFC method, characterized by the average relative deviation of PCFC results from the corresponding oxygen bomb calorimetry measurements, is about 3 %.

Discussion

The method described in this paper separately reproduces the condensed phase (pyrolysis) and gas phase (combustion) processes of flaming combustion in a single laboratory test. Decoupling the condensed and gas phase processes in this way and forcing them to completion isolates the thermal chemistry of the condensed phase and allows detailed investigation of the relationship between chemical structure and the amount, rate and temperatures of heat release under fire-like conditions [20–26,59–61,63]. The use of small (milligram) samples to maintain thermal equilibrium during the present test ensures that the results are insensitive to thermal-physical phenomena such as melting, dripping, swelling, shrinking, and char layer growth as well as extrinsic factors such as sample size and orientation, boundary conditions, venti-

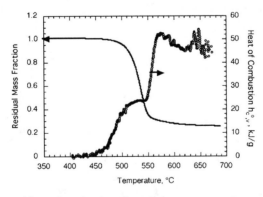

FIG. 15—*Residual mass and heat of combustion of pyrolysis gases versus temperature for test of polycar-bonate in TGA at* $\beta = 20$ *K/min.*

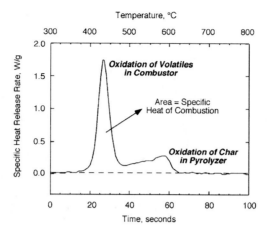

FIG. 16—*Specific heat release rate of polycarbonate versus time and temperature obtained by oxidative pyrolysis-combustion. Oxidation of gases in combustor and solid char in pyrolyzer are seen as separate processes.*

lation rate, etc.—all of which influence flame and fire test results of large samples. Despite the fact that the present test method measures the maximum potential (capacity) of the material to release heat in flaming combustion (η_c), it is a reasonable predictor of flame and fire test results [25,26].

Conclusions

An apparatus and methodology is presented that uses controlled heating of milligram samples and complete combustion of the evolved gases to separately reproduce the condensed phase (pyrolysis) and gas phase (combustion) processes of flaming combustion in a single laboratory test. A simple combustion model explains the test results including the amount, rate and temperatures of heat released under fire-like conditions. A flammability parameter is measured that is the maximum potential (capacity) of a material to release heat in flaming combustion.

Acknowledgments

The authors are indebted to Lauren Castelli and Qaadir Williams for the experimental values of heat release capacity. Although certain commercial equipment, instruments, materials and companies are identified in this paper in order to adequately specify the experimental procedure, this in no way implies endorsement or recommendation by the Federal Aviation Administration.

TABLE 4—*Net heat of combustion of charring ($\mu \neq 0$) and noncharring ($\mu = 0$) polymers obtained by oxygen bomb calorimetry and oxidative pyrolysis-combustion flow calorimetry (oPCFC). (Triplicate determinations).*

Polymer (μ, kg/kg)	ASTM D 2015 (MJ/kg)	oPCFC (MJ/kg)	Relative Deviation (%)
Polyethylene (0)	43.3	43.5	0.5
Polystyrene (0)	39.8	39.4	−1.0
Polymethylmethacrylate (0)	24.9	25.0	0.4
Polyoxymethylene (0)	15.9	16.0	0.6
Polybutyleneterephthalate (0.02)	26.7	26.3	−1.5
Polyethyleneterephthalate (0.13)	21.8	23.2	6.4
Polycarbonate (0.23)	29.8	29.1	−2.3
Polyaramide fiber (0.36)	27.8	28.1	1.1
Polyetheretherketone (0.47)	30.2	30.9	2.3
Phenolic Triazine (0.67)	29.8	29.5	−1.0

References

[1]   Pearce, E. M., Khanna, Y. P., and Raucher, D., "Thermal Analysis of Polymer Flammability," in *Thermal Characterization of Polymeric Materials*, E. A. Turi, Ed., Academic Press, Orlando, FL, 1981, pp. 793–843.

[2]   Hassel, H. L., "Evaluation of Polymer Flammability by Thermal Analysis," American Laboratory, 9(1), 1977, pp. 35–37.

[3]   Aseeva, R. A. and Grygorovskaya, V. A., "Thermal Properties and Flammability of Polyarylenes and their Modifying Products," *Polym. Degrad. Stab.*, Vol. 64, pp. 457–463 (1999).

[4]   Mark, H. F., Atlas, S. M., Shalaby, S. W., and Pearce, E. M., "Combustion of Polymers and its Relatardation," in *Flame Retardancy of Polymeric Materials*, M. Lewin, S. M. Atlas, and E. M. Pearch, Eds., Plenum Press, NY, 1975, pp. 1–17.

[5]   Cullis, C. F and Hirschler, M. M, "The Significance of Thermoanalytical Measurements in the Assessment of Polymer Flammability," *Polymer*, Vol. 24, pp. 834–840 (1983). .

[6]   Weisner, E., "Study of the Flammability of Polymers by Thermal Analysis," Chemicke Vlakna, 26(3-4), 1976, pp. 146–150.

[7]   Mickelson, R. W., "Using Thermal Analysis to Confirm the Flammability Behavior of a Polymer," *Proceedings of the American Chemical Society Meeting*, 33(1), 1973, pp. 485–489.

[8]   Carroll-Porczynski, C. Z., "Applications of Simultaneous DTA/TGA and DTA/MS Analysis for Predicting the Flammability of Composite Textile Fabrics and Polymers," *Composites*, 4(1), 1973, pp. 9–15.

[9]   Cullis, C. F., "The Role of Pyrolysis in Polymer Combustion and Flame Retardance," *J. Anal. Appl. Pyrolysis*, 11, 1987, pp. 451–463.

[10] Balog, K., Kosik, S., Kosik, M., Reiser, V., and Simek, I., "Application of Thermal Analysis Procedures to the Combustion and Flammability of Some Polymers," *Thermochim. Acta*, 93, 1985, pp. 167–170.

[11] Simon, J., Androsits, B., and Kozma, T., "Study of the Process of Flame Retardation by Thermal Analysis," Magyar Kemikusok Lapja, 36(10), 1981, pp. 510–514.

[12] van Krevelen, D. W., "Some Basic Aspects of Flame Resistance of Polymeric Materials," *Polymer*, Vol. 16, 1975, pp 615–620.

[13] Carty, P. and White, W., "The Importance of Char Forming Reactions in Thermoplastic Polymers," *Fire Mater.*, 18, 1994, pp. 151–166.

[14] Murashko, E. A., Levchik, G. F., Levchik, S. V., Bright, D. A., and Dashevsky, S., "Fire Retardant Action of Resorcinol Bis(Diphenyl Phosphate) in a PC/ABS Blend. I. Combustion Performance and Thermal Decomposition Behavior," *J. Fire Sci.*, Vol. 16, 1998, pp. 278–295.

[15] Hirschler, M. M., "Chemical Aspects of Thermal Decomposition of Polymeric Materials," in *Fire Retardancy of Polymeric Materials*, A. F. Grand and C. A. Wilkie, Eds., Marcel Dekker, New York, 2000, pp. 28–75.

[16] Gracik, T. D. and Long, G. L., "Prediction of Thermoplastic Flammability by Thermo-gravimetry," *Thermochim. Acta*, 212, 1992, pp. 163–170.

[17] Kashiwagi, T., "Polymer Combustion and Flammability-Role of the Condensed Phase," *Proceedings of the 25th Symposium (International) on Combustion*, The Combustion Institute, 1994, pp. 1423–1437.

[18] Lyon, R. E., "Solid-State Thermochemistry of Flaming Combustion," in: *Fire Retardancy of Polymeric Materials*, A. F. Grand and C. A. Wilkie, Eds., Marcel Dekker, Inc., New York, NY, 2000, pp. 391–447.

[19] Lyon, R. E., 'Heat Release Kinetics," *Fire Mater.*, 24, 2000, pp. 179–186.

[20] Lyon, R. E., "Heat Release Capacity," *Proceedings of the Fire & Materials Conference*, San Francisco, CA, January 22–24, 2001.

[21] Lyon, R. E., "A New Method for Measuring Polymer Flammability," *Flame Retardants 2002*, London, England, February 5–6, 2002.

[22] Lyon, R. E. and Walters, R. N., "Heat Release Capacity: A Molecular Level Fire Response Parameter," *Proceedings of the 7th International Symposium on Fire Safety Science*, Intl. Assoc. for Fire Safety Science, Worcester, MA, 2003, pp. 1167–1168.

[23] Walters, R. N. and Lyon, R. E., "Molar Group Contributions to Polymer Flammability," *J. Appl. Polym. Sci.*, 87, 2003, pp. 548–563.

[24] Zhang, H., "Fire Safe Polymers and Polymer Composites," Final Report DOT/FAA/AR-04/11, September 2004.

[25] Lyon, R. E., "Plastics and Rubber," in *Handbook of Building Materials for Fire Protection*, C. A. Harper, Ed., McGraw-Hill, New York, NY, Chapter 3, 2004, pp. 3.1–3.51.

[26] Lyon, R. E. and Janssens, M., "Polymer Flammability," *Encyclopedia of Polymer Science & Engineering* (on-line edition), John Wiley & Sons, New York, NY, October 2005.

[27] Lyon, R. E., "Pyrolysis Kinetics of Char Forming Polymers," *Polym. Degrad. Stab.*, 61(2), 1998, pp. 201–210.

[28] Moghtaderi, B., Novozhilov, V., Fletcher, D., and Kent, J. H., "An Integral Model for the Transient Pyrolysis of Solid Materials," *Fire Mater.*, 21, 1997, pp. 7–16.

[29] Quintiere, J. and Iqbal, N., "An Approximate Integral Model for the Burning Rate of a Thermoplastic-Like Material," *Fire Mater.*, 18, 1994, pp. 89–98.

[30] Chen, Y., Delichatsios, M. A., and Motevalli, V., "Material Pyrolysis Properties, Part I: An Integral Model for One-Dimensional Transient Pyrolysis of Charring and Non-Charring Materials," *Combust. Sci. Technol.*, 88, 1993, pp. 309–328.

[31] Staggs, J. E. J. and Whiteley, R. H., "Modeling the Combustion of Solid-Phase Fuels in Cone Calorimeter Experiments," *Fire Mater.*, 23, 1999, pp. 63–69.

[32] Staggs, J. E. J. and Nelson, M. I., "A Critical Mass Flux Model for the Flammability of Thermoplastics," *Combust. Theory Modell.*, Vol. 5, 2001, pp. 399–427.

[33] Staggs, J. E. J., "Modeling Thermal Degradation of Polymers Using Single-Step First-Order Kinetics," *Fire Saf. J.*, 32, 1999, pp. 17–34.

[34] Bucsi, A. and Rychly, J., "A Theoretical Approach To Understanding The Connection Between Ignitability And Flammability Parameters Of Organic Polymers," *Polym. Degrad. Stab.*, 38, 1992, pp. 33–40.

[35] Lyon, R. E. "An Integral Method of Nonisothermal Kinetic Analysis," *Thermochim. Acta*, 297, 1997, pp. 117–124.

[36] Grassie, N. and Scotney, A., "Activation Energies for Thermal Degradation of Polymers," in *Polymer Handbook*, 2nd ed., J. Brandrup and H. Immergut, Eds., Wiley-Interscience, NY, 1975, pp. II. pp. 467–471.

[37] Van Krevelen, D. W., *Properties of Polymers*, Elsevier, Amsterdam, 1990, pp. 627–653.

[38] Bicerano, J., *Prediction of Polymer Properties*, 2nd ed., Marcel Dekker, New York, NY, 1996.

[39] ASTM Standard E 2008, Test Method for Volatility Rate by Thermogravimetry, ASTM International, West Conshohocken, PA.

[40] ASTM Standard E 1641, Test Method for Decomposition Kinetics by Thermogravimetry, ASTM International, West Conshohocken, PA.

[41] Walters, R. N., "Molar Group Contributions to the Heat of Combustion," *Fire Mater.*, 26, 2002, pp. 131–145.

[42] Babrauskas, V., "Heat of Combustion and Potential Heat," in: V. Babrauskas and S. J. Grayson, Eds., *Heat Release in Fires* (Chapter 8), Elsevier, New York, 1992, pp. 207–223.

[43] Thornton, W. M., "The Relation of Oxygen to the Heat of Combustion of Organic Compounds," *Philos. Mag.* 33, 1917, p. 196.

[44] Huggett, C., "Estimation of Rate of Heat Release by Means of Oxygen Consumption Measurements," *Fire Mater.*, 4(2), 1980, p. 61.

[45] Janssens, M. L., "Measuring Heat Release by Oxygen Consumption," *Fire Technol.*, 27, 1991, p. 234.

[46] Janssens, M. and Parker, W. J., "Oxygen Consumption Calorimetry," in: V. Babrauskas and S. J. Grayson, Eds., *Heat Release in Fires* (Chapter 3), Elsevier, New York, 1992, pp. 31–59.

[47] ASTM Standard 1354, Test Method for Heat and Visible Smoke Release Rates for Materials and Products Using and Oxygen Consumption Calorimeter, ASTM International, West Conshohocken, PA.

[48] ASTM Standard E 2058, Test Method for Measurement of Synthetic Polymer Material Flammability Using a Fire Propagation Apparatus (FPA), ASTM International, West Conshohocken, PA.

[49] Walters, R. N., Hackett, S. M., and Lyon, R. E., "Heats of Combustion of High Temperature Polymers," *Fire Mater.*, 24, 2000, pp. 245–252.

[50] ASTM Standard D 2015, Test Method for Gross Calorific Value of Coal and Coke by the Adiabatic Bomb Calorimeter, ASTM International, West Conshohocken, PA.

[51] Lyon, R. E. and Walters, R. N., "A Pyrolysis-Combustion Flow Calorimeter for the Study of Polymer Heat Release Rate," *Proceedings, 9th Annual BCC Conference on Flame Retardancy of Polymeric Materials*, Stamford, CT, June 1–3, 1998.

[52] Microscale Combustion Calorimeter, U.S. Patent 5,981,290, November 9, 1999.

[53] Lyon, R. E. and Walters, R. N., "A Microscale Combustion Calorimeter," Final Report DOT/FAA/AR-01/117, February 2002.

[54] Heat Release Rate Calorimeter for Milligram Samples, U.S. Patent 6,464,391, October 15, 2002.

[55] Lyon, R. E. and Walters, R. N., "Pyrolysis Combustion Flow Calorimetry," *J. Anal. Appl. Pyrolysis*, 71(1), 2004, pp. 27–46.

[56] Heffington, W. M., Parks, G. E., Sulzmann, K. G. P., and Penner, S. S., "Studies of Methane Oxidation Kinetics," *Sixteenth Symposium (International) on Combustion*, The Combustion Institute, 1976, pp. 997–1010.

[57] Reshetnikov, S. M. and Reshetnikov, I. S., "Oxidation Kinetic of Volatile Polymer Degradation Products," *Polym. Degrad. Stab.*, 64, 1999, pp. 379–385.

[58] Babrauskas, V., Parker, W. J., Mulholland, G., and Twilley, W. H., "The Phi Meter: A Simple, Fuel-Independent Instrument for Monitoring Combustion Equivalence Ratio," *Rev. Sci. Instrum.*, 65(7), 1994, pp. 2367–2375.

[59] Schoemann, A., Westmoreland, P. R., Zhang, H., Farris, R. J., Walters, R. N., and Lyon, R. E., "A Pyrolysis/GC-MS Method for Characterizing Flammability and Thermal Decomposition of Polymers," *Proceedings 4th Joint Meeting of the U.S. Sections of the Combustion Institute*, Philadelphia, PA, March 21-23, 2005.

[60] Westmoreland, P. R., Inguilzian, T., and Rotem, K., "Flammability Kinetics from TGA/DSC/GCMS, Microcalorimetry and Computational Quantum Chemistry," *Thermochim. Acta* 67, 2001, pp. 401–405.

[61] Inguilizian, T. V., "Correlating Polymer Flammability Using Measured Pyrolysis Kinetics," Master of Science Thesis, University of Massachusetts, Amherst, January 1999.

[62] Factor, A., "Char Formation in Aromatic Engineering Polymers," *Fire and Polymers*, G. L. Nelson, ed., ACS Symposium Series 425, American Chemical Society, Washington, DC, 1990, pp. 274–287.

[63] Hergenrother, P. M, Thompson, C M, Smith, J G, Jr, Connell, J. W, Hinkley, J. A, Lyon, R. E, and Moulton, R, "Flame Retardant Aircraft Epoxy Resins Containing Phosphorus," *Polymer* Vol. 46, 2005, pp. 5012–5024.

Journal of ASTM International, Vol. 4, No. 3
Paper ID JAI100528
Available online at www.astm.org

Alan T. Riga,[1,3] *Michael Golinar,*[2] *and Kenneth S. Alexander*[3]

Fast Scan Differential Scanning Calorimetry Distinguishes Melting, Melting-Degradation/Sublimation and Thermal Stability of Drugs

ABSTRACT: In order to establish a structure and property (melting and oxidative or thermal degradation, or both) relationship for a United States Pharmacopeias (USP) set of standard drugs, they were evaluated by fast scan differential scanning calorimetry. A critical problem in characterizing the endothermic melting of a drug is to determine the melting range and if a chemical melts and immediately degrades. The stability of standard drugs is based on a comparison of their thermal properties at widely varying ramp or heating rates from 10 to 100°C/min. A stable crystalline drug has an obvious melting endotherm followed by a stable baseline. An unstable crystalline drug melts and immediately degrades as viewed by a shifting melt endotherm with heating rate. The USP thermally stable standards evaluated in this study include vanillin (melt temperature, T_m, 80.4°C), acetanilide (T_m, 114°C), acetophenetidin (T_m, 135°C), sulfanilamide (T_m, 165°C), sulfapyridine (T_m, 191°C), and caffeine (T_m, 235°C and $T_{sublimation}$, <220°C). In addition to the USP samples a number of commercial and model drugs, like benzoic acid (T_m, 122°C and $T_{sublimation}$, <120°C), lidocaine-.HCl and procaine.HCl were also examined. Their melt profiles were ranked as stable or unstable post fusion by the fast scan DSC technique and are reported.

Introduction

An important innovation of differential scanning calorimetery (DSC) is fast scan DSC (FSDSC) where the sample under standard operating conditions [1] examined at a heating rate or ramp of 10°C/min is now examined with a ramp of 100 to 500°C/min [2]. These high ramp rates mimic processing conditions. This new advanced DSC technique "FSDSC" has impacted material science by allowing the measurement of a great number of physical and chemical properties at high ramp (heating rate) that are time independent separating them from time dependent phenomena. Some of these events that have been minimized or eliminated are transitions for various metastable forms of polymorphic drugs and improvement of drug stability rendering improved baselines for more quantitative analysis. Melting and crystallization transitions as well as enthalpy of fusion and crystallization are more clearly denoted at the higher heating rates. Modern DSC instrumentation has yielded stable and quality sensors allowing this advanced technique to become part of the arsenal of thermal analytical methods in characterizing materials, especially drugs and polymers [3].

Cassel and Wiese reported that FSDSC quantified metastable forms in polymers and pharmaceuticals determining potentially useful structural information [4]. The fast rate method was found to be more accurate in determining the initial crystallinity in a polymer, e.g., polyethylene terephthalate. The FSDSC was identified as being more capable of determining specific heat data that were more representative of the material in its original state. The high heat rates also inhibited crystallization and reorganizaton during heating which produced higher melting more stable crystals. The controlled quench cooled (100°C/min) crystalline polymer from the melt was more representative of the desired crystallite structure. It was pointed out that the FSDSC pharmaceutical applications aids in characterizing target drugs and excipients polymorphic forms. Analysis of the metastable forms of drugs assists the preformulation development of

---

Manuscript received March 7, 2006; accepted for publication February 15, 2007; published online April 2007. Presented at ASTM Symposium on Techniques in Thermal Analysis: Hyphenated Techniques, Thermal Analysis of the Surface, and Fast Rate Analysis on 24 May 2005 in West Conshohocken, PA; L. Judovits and W.-P. Pan, Guest Editors.
[1] Department of Chemistry, 2121 Euclid Avenue, SI 329, Cleveland State University, Cleveland OH 44115-2406, e-mail: alanriga@sbcglobal.net
[2] TA Instruments, 109 Lukens Drive, New Castle, DE 19720.
[3] College of Pharmacy, University of Toledo, Toledo, OH 43606-3390.

TABLE 1—*Summary of effect of ramp on melting and Heat of Fusion: average $T_m(To/e)$, T_p, and ΔH.*

Drug	Ramp (°C/min)	DSC Melting To/e (°C)	T_p (°C)	ΔH (J/g)	References T_m(lit) (10-18)	To/e (°C)	T_p (°C)	ΔH (J/g)
Acetophenetidin	10	135	137	140				
	100	138	143	170		Asymmetric peak		
average		136	140	155	135			
Benzoic Acid	10	122	125	129				
	100	123	124	127		Sublimation		
average		122	124	128	122			
Caffeine	10 cp	236	237	98		154	159	17
	50 cp	236	238	101		156	164	20
	75	236	238	88		Sublimation		
	100	236	236	93		157	164	16
average		236	237	95	235	156	162	18
Lidocaine.HCl	10	74	81	138				
	75	84	89	144				
	100	84	88	135				
average		81	86	139	74–79			
Acetanilide	10 cp	114	116	154				
	75	114	115	156		Asymmetric peak		
average		114	116	155	114			
Vanillin	10 cp	81	84	149				
	100	82	85	140		Asymmetric peak		
average		82	84	144	80			
Sulfapyridine	10 cp	191	193	154				
	75	192	192	150				
	100	191	193	150				
average		191	193	151	191			
Procaineamide.HCl	100	170	172		165–169			
Sulfanilamide	100	164	165		165			

Notes: T_m Literature References=Average % relative error. CAS reference numbers cited in Table 2 for top seven drugs cited above=1.5%.

formulated drugs. This study yields data on constructing phase diagrams and predicting the relative stability of polymorphs as a function of temperature. In summary, this technique is clearly a new tool for structural clarification. Small quantity samples can be routinely analyzed with improved throughput and analysis [4].

Gabbott et al. [5] reported on "Hyper DSC in Pharmaceuticals." This technique simplifies identification of the glass transition in amorphous materials, e.g., the amorphous excipient, lactose. A spray dried lactose that was almost 100% amorphous content was analyzed at heating rates from 20 to 500°C/min. The T_g was in the range of 100 to 120°C. A lower lactose T_g is attributed to a plasticized lactose. Hyper DSC with increased sensitivity clearly revealed the T_g in the range of 80 to 100°C. Plasticization water is not lost during Hyper DSC and therefore an accurate measure of the T_g is available. Following the T_g there was an exotherm associated with recrytallization then melting. It was also reported that Hyper DSC can be used as a pharmaceutical screening tool. A sample of Dotriacontane was examined at 250°C/min and employing the second derivative clearly rendered several transitions which are not viewed in conventional DSC.

It is our definition that in FSDSC, the difference in energy input into a substance and a reference material is measured as a function of temperature, while the substance and reference material are subjected to a controlled temperature program at rates exceeding 100°C/min [6].

The focus of this study was to determine the standard DSC melt profile of United States Pharmacopoeia (USP) drugs, that is, their melt temperature and heat of fusion under conditions described in ASTM E 793, E 967, and E 968 [7–9]. These ASTM protocols call for ramping DSC temperature at 10°C/min in nitrogen by conventional DSC. The overall DSC focus was on baseline stability to best determine the heat of fusion and melt temperatures. In this study the ramp was widely varied from 10 to 100°C/min as was

TABLE 2—*DSC melting profile of drugs.*

	Drug	pan	Ramp (°C/min)	T_m (°C)	T_p (°C)	ΔH (J/g)	Figure	Molecular Formula Literature Reference No.	Reference
1	Acetophenetidin	open	10	135	137	140	1	$CH_3CONHC_6H_4OC_2H_5$	[10]
		open	100	138	143	170		CAS No: 62-44-2	
2	Benzoic acid	open	10	122	125	129	2	C_6H_5COOH	
		open	100	123	124	127		CAS No. 65-85-0	[11]
3	Caffeine	closed	10	236	237	98	3	$C_8H_{10}N_4O_2$	
		closed	50	236	238	101		CAS No: 58-08-2	[12]
		open	75	236	238	88			
		open	100	236	236	93			
4	Acetanilide	closed	10	114	116	154	4	$CH_3CONHC_6H_5$	
		open	75	114	115	157		CAS No: 103-84-4	[13]
5	Sulfapyridine	closed	10	191	193	154	5	$C_{11}H_{11}N_3O_2S$	
		open	75	192	193	150		CAS No: 144-83-2	[14]
		open	100	192	193	150			
6	Lidocaine.HCl	closed	10	74	81	138	6	$C_{14}H_{22}N_2O$.HCl	
		open	75	84	89	144		CAS No: 137-58-6	[15]
		open	100	84	88	135			
7	Vanillin	closed	10	81	84	149	7	$C_8H_8O_3$	
		open	100	82	85	140		CAS No: 121-33-5	[16]
8	Procainamide.HCl	open	100	170	172		no Fig.	$C_{13}H_{21}N_3O$.HCl	
								CAS No: 51-06-9	[17]
9	Sulfanilamide	open	100	164	165		no Fig.	$C_6H_8N_2O_2S$	
								CAS No: 83-74-1	[18]

Key: T_m=Extrapolated onset melt temperature (°C). T_p=Peak temperature (°C). ΔH=Heat of fusion (J/g).

the pan configuration and the effect of these variables was determined on melt, sublimation, and degradation profiles. Next we examined several commercial drugs with the new FSDSC protocol at 100°C/min and compared the results to those observed for 10°C/min. We wanted to minimize other thermal events, for example, solid-solid transitions or sublimation by varying the ramp as a tool.

Experimental Samples and Procedures

The following USP samples were examined in this investigation: vanillin, T_m, 80.4°C; acetanilide, T_m, 114°C; acetophenetidin, T_m, 135°C; sulfanilamide, T_m, 165°C; sulfapyridine, T_m, 191°C; caffeine, T_m, 235°C and sublimation temperature, <220°C); and benzoic acid, T_m, 122°C and sublimation temperature, <120°C. We also examined lidocaine.HCl and procainamide.HCl by this method.

The TAI 2920 Modulated DSC in a cool-heat-cool mode evaluated 2–5 mg of the drugs at 10–50°C/min in N_2 with a closed crimped aluminum pan. The TAI robotic Q1000 in a cool-heat-cool mode evaluated ca. 5 mg of the drugs at 75 and 100°C/min in N_2 in an open aluminum pan. The DSC under fast scan conditions was calibrated at each heating rate in this study prior to testing of the drug samples

Results and Discussion

A summary of the effect of ramp on melting temperature and heat of fusion for the nine drugs or excipients is cited in Tables 1 and 2. These tables describes the drug, ramp (°C/min), DSC melting onset temperature (melting temperature, T_m, or To/e, extrapolated onset temperature, °C), melting peak temperature T_p (°C), Delta H (Heat of Fusion, J/g), T_m (literature values from the ACS Chemical Abstracts, [10–18]), and Transition Properties as To/e, T_p and Delta H (J/g). Some ramp rates were designated as "cp" which means that the aluminum pan was crimped closed while all the others were open during the DSC examination. A comparison of the average melting temperature, T_m, for each sample examined under varying conditions is boxed in Table 1. The average melting temperature percent relative error for seven drugs is 1.5 %, a valued statistical assessment. Therefore the T_m did not vary with heating rate. However, the Heat of Fusion (ΔH) for example, acetophenetidin varied from 140 to 170 J/g or the average was 155±15 J/g or ±9.7 %. The DSC curve of acetophenetidin is Fig. 1.

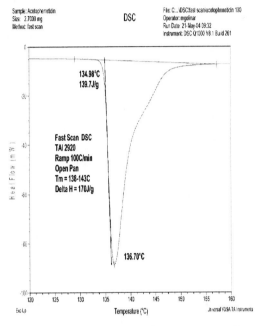

FIG. 1—*DSC of acetophenetidin; data summarized in Tables 1 and 2.*

Benzoic acid DSC curve is Fig. 2 and the baseline above the melting clearly indicates an additional phenomena is occurring. In this case benzoic acid is melting and subliming.

The T_m, T_p, and ΔH (J/g) did not vary with the DSC experimental conditions. The known T_m is 122 and the measured T_m is 122°C. Again no variation in melting properties with ramp is noted. The benzoic acid sublimation did not interfere in collecting the appropriate DSC data.

Caffeine is noted for subliming when heated. The T_m and T_p showed no difference with the open aluminum pan (75 and 100°C/min) and the closed crimped aluminum pan (10 and 50°C/min), see Fig. 3. The observed and literature T_m were within 1 °C. The heat of fusion did vary with the average overall at 95±7 J/g or ±7.4 %. The closed pan heat of fusion varied by 2 % and the open pan by 3 %. Therefore,

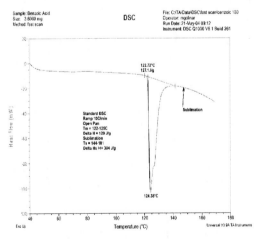

FIG. 2—*DSC of benzoic acid; data summarized in Tables 1 and 2.*

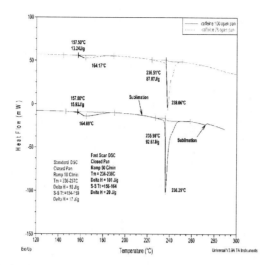

FIG. 3—*DSC of caffeine at varying heating rates. Data summarized in Tables 1 and 2.*

the open pan heat evaluation with enhanced sublimation was repeatable but caused a marked variation from the closed pan heat of fusion. The FSDSC ramp minimized the sublimation as evidenced by only a slight endothermic bend in the curve as compared to a significant bend in the DSC curve for benzoic acid, compare Figs. 2 and 3.

The DSC of acetanilide at 10°C/min (closed cup) and 75°C/min (open cup) are reported in Fig. 4, Tables 1 and 2. The T_m, T_p, and Heat of Fusion showed little or no variation. The average T_m observed at 114°C was identical to 114°C in the literature. The Heat of Fusion variation was 155±1 J/g or ±0.6 %. The FSDSC did render an asymmetric melting peak for acetanilide.

The DSC of sulfapyridine at various heating rates is recorded in Fig. 5, Tables 1 and 2. The T_m and T_p did not vary for heating rates of 10°C/min (closed cup), 75 (open cup), and 100°C/min (open cup). The overall Heat of Fusion was 151±3 J/g or ±2 %. The open and closed cup heat varied by 0.6 %. The 100°C/min peak was slightly asymmetric.

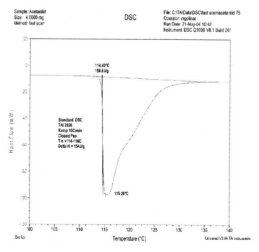

FIG. 4—*DSC of acetanilide at 10 and 75°C/min; data summarized in Tables 1 and 2.*

Fast Scan DSC of Sulfapyridine

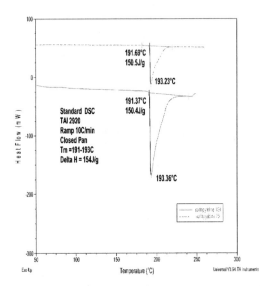

FIG. 5—*DSC of sulfapyridine at various heating rates; data summarized in Tables 1 and 2.*

The DSC of lidocaine.HCl, a commercial analgesic, was evaluated in the FSDSC protocol and reported in Fig. 6, Tables 1 and 2. The average T_m 81°C was high for the T_m reported in the literature at 74–79°C. However, the sample examined in this study was the HCl salt of lidocaine and probably contributed to the variation. There was a 10°C difference in T_m at 10°C/min (closed cup) versus the 75 and 100°C/min (open cup); the widest variation noted in this study. The average Heat of Fusion was

Fast Scan DSC of Lidocaine at Various Heating Rates

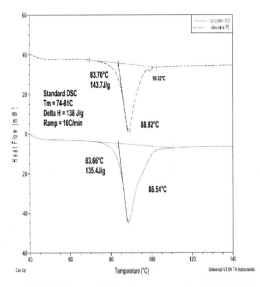

FIG. 6—*DSC of lidocaine.HCl; data summarized in Tables 1 and 2.*

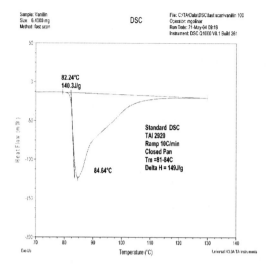

FIG. 7—*DSC of vanillin at 100°C/min; all data summarized in Tables 1 and 2.*

139±5 J/g or 3.5 % maximum. It does not appear that the open or closed cup caused the changes noted in the lidocaine.HCl thermal data.

Vanillin, an excipient, examined with this DSC protocol rendered a T_m (82°C) and T_p (84°C) with little or no variation, Fig. 7. The average T_m was 2°C higher than the T_m reported in the literature. The Heat of Fusion was 144±4 J/g or ±2.8 %. The latter could be related to a higher value in the closed cup with a better thermal contact and a lower value in the open cup.

Procaineamide.HCl and sulfanilamide at 100°C/min are summarized in Tables 1 and 2. The USP sulfanilamide T_m was in good agreement with the literature value. The commercial procaineamide.HCl showed a similar variation as lidocaine.HCl with a T_m 3–5°C higher as determined at 100°C/min than the literature value.

Conclusions

The FSDSC technique adds significantly to the DSC methods to enhance the characterization of materials and especially polymers, drugs, and excipients. When evaluating a drug or excipient by the FSDSC method heating rate and pan configuration caused little or no variation in melting temperature and heat of fusion values. Productivity in the drug testing and analysis lab can be greatly enhanced by employing heating rates of 100–500°C/min with confidence that precision and accuracy are maintained. However, one must know the material under evaluation before employing this technique as we discovered by studying the commercial drugs procaine and lidocane.

References

[1] Riga, A., and Collins, R., "Differential Scanning Calorimetry and Differential Thermal Analysis," *Encyclopedia of Analytical Chemistry*, R. A. Meyers, Ed., John Wiley, Chichester, UK, 2000, pp. 13147–13179.

[2] Thomas, L., Short Course Instructor, NATAS Conference, Albuquerque, NM, September 2003.

[3] Sauerbrunn, S., Short Course Instructor, NATAS Conference, Albuquerque, NM, September 2003.

[4] Cassel, B., and Wiese, M., "Fast Scan DSC Quantifies Metastable Forms in Polymers and Pharmaceuticals," *American Laboratory*, International Scientific Communications, Inc., January 2003, pp. 13–16.

[5] Gabbott, P., Clarke, P., Mann, T., Royall, P., and Shergill, S., "A High Sensitivity High Speed DSC Technique: Measurement of Amorphous Lactose," *American Laboratory*, International Scientific

Communications, Inc., August 2003, pp. 17–22.

[6] Riga, A., Golinar, M., and Alexander, K., "FSDSC of Drugs and Excipients," *ASTM E 37 Symposium*, March 2004.

[7] ASTM, *Standard E 0793, Annual Book of ASTM Standard* Vol. 14.02, ASTM, West Conshohocken, PA, 2006.

[8] ASTM, *Standard E 0967, Annual Book of ASTM Standard* Vol. 14.02, ASTM International, West Conshohocken, PA, 2006.

[9] ASTM, *Standard E 0968, Annual Book of ASTM Standard* Vol. 14.02, ASTM International, West Conshohocken, PA, 2006.

[10] CAS no. 62-44-2, Acetophenetidin.

[11] CAS no. 65-85-0, Benzoic Acid.

[12] CAS no. 58-08-2, Caffeine.

[13] CAS no. 103-84-4, Acetanilide.

[14] CAS no. 144-83-2, Sulfapyridine.

[15] CAS no. 137-58-6, Lidocaine.HCl.

[16] CAS no. 121-33-5, Vanillin.

[17] CAS no. 51-06-9, Procaineamide.HCl.

[18] CAS no. 83-74-1, Sulfanilamide.

Journal of ASTM International, Vol. 4, No. 3
Paper ID JAI100547
Available online at www.astm.org

Alan T. Riga,[1,2,3] *Kenneth S. Alexander,*[2] *and Kevin Williams*[3]

Thermal and Oxidative Properties of Physiologically Relevant Free Fatty Acids by Dielectric Analysis and Differential Scanning Calorimetry

ABSTRACT: Physiologically relevant fatty acids and related organic acids are basic for human life. The essential fatty acids, linoleic, linolenic, and arachidonic acids, are sourced from vegetable seed oils (corn, sunflower, safflower), and margarines blended with vegetable oils. The functions of these special acids are in the synthesis of prostaglandins and membrane structures. Growth cessation and dermatitis occurs with a deficiency of the fatty acids. A typical therapeutic dosage of the essential fatty acids is up to 10 g per day. The polyunsaturated fatty acids. linoleic (9,12-octadecaidienoic), linolenic (9,12,15-octadecatrienoic), and arachidonic (5,8,11,14-eicosatetraenoic) are referred to as essential fatty acids. They unlike other lipids must be provided by diet. Arachidonic acid can be produced in the body by linoleic acid. This thermal analytical study is to determine fatty acids' physical transitions [melting] by DSC at low temperatures and their surface properties by low frequency dielectric analysis and relate those properties to the inherent amount of unsaturation in the fatty acids. It is our premise that the degree of unsaturation will affect low temperature melt temperature and electrical properties, e.g., electrical conductivity and complex permittivity. We have observed that the DEA properties of the air-aged liquid fatty acids indicate that the electrical conductivity and complex permittivity can be correlated with the degree of unsaturation. It is our objective to establish a relationship between the amount of unsaturation, number of double bond sites and the electrical properties, complex permittivity, and electrical conductivity.

KEYWORDS: polyunsaturated fatty acids, linoleic (9,12-octadecaidienoic), linolenic (9,12,15-octadecatrienoic), arachidonic (5,8,11,14-eicosatetraenoic), oxidative behavior of fatty acids, dielectric analysis

Introduction

Physiologically relevant fatty acids and related organic acids are basic for human life. The essential fatty acids, linoleic, linolenic, and arachidonic acids, are sourced from vegetable seed oils (corn, sunflower, safflower), and margarines blended with vegetable oils. The functions of these special acids are in the synthesis of prostaglandins and membrane structures. Growth cessation and dermatitis occurs with a deficiency of the fatty acids. A typical therapeutic dosage of the essential fatty acids is up to 10 g per day. The polyunsaturated fatty acids, linoleic (9,12-octadecaidienoic), linolenic (9,12,15-octadecatrienoic), and arachidonic (5,8,11,14-eicosatetraenoic) are referred to as essential fatty acids. They unlike other lipids must be provided by diet. Arachidonic acid can be produced in the body by linoleic acid.

The electron transfer oxidation properties of unsaturated fatty acids were studied and gave light to the mechanistic insight into lipoxygenases [1]. This study revealed a one electron oxidation potential of unsaturated fatty acids and described the intrinsic barrier to electron transfer. The electron transfer rate constant of linoleic acid, linolenic acid, and arachidonic acid were similar leading to the same oxidation potential of 1.76 V versus SCE. This potential was significantly lower than that of oleic acid (2.03 V versus SCE). This research provided valuable insight into the mechanism of lipoxygenases which followed a proton-coupled electron transfer process during the catalytic process.

The three essential fatty acids are precursors of prostaglandins and vitamin B_6 and are involved in their metabolism. Therefore to understand the basics, nature, and mechanism of the oxidation of prostaglandins,

Manuscript received March 17, 2006; accepted for publication February 15, 2007; published online April 2007. Presented at ASTM Symposium on Techniques in Thermal Analysis: Hyphenated Techniques, Thermal Analysis of the Surface, and Fast Rate Analysis on 24 May 2004 in West Conshohocken, PA; L. Judovits and W.-P. Pan, Guest Editors.
[1] Department of Clinical Chemistry, Cleveland State University, Cleveland, OH 44115.
[2] College of Pharmacy Practice, University of Toledo, Toledo, OH 43606.
[3] Department of Chemistry, Western Kentucky University, Bowling Green, KY 42101.

vitamins, and triglycerides, we undertook the study of the physical properties of the essential fatty acids.

A purpose of this study is to determine the free fatty acids' physical transitions such as melting and crystallization by thermal analytical methods. It is our hypotheses that the variation in the acid structure with accompanying amount of unsaturation (number of double bonds) has a pronounced effect on the chemical's physical properties, for example, melt temperature, electrical conductivity, and oxidation. The fusion temperature, heat of fusion, crystallization temperature, and heat of crystallization will vary with structural variations proportional to the chemical's number of double bonds. Accompanying and supporting the structure property relation is the measurement of these acids' dielectric analysis properties, as the permittivity or complex permittivity, loss factor or ac electrical conductivity and tan delta (ratio of loss factor by permittivity). A major focus of this study is to evaluate the essential fatty acids by dielectric analysis (DEA) as a function of temperature and at low frequencies (0.10 Hz to 50 Hz).

In summary, this thermal analytical study is to determine fatty acids' physical transitions [melting] by DSC at low temperatures and their surface properties by low frequency DEA and relate those properties to the inherent amount of unsaturation in the fatty acids. It is our premise that the degree of unsaturation will affect low temperature melt temperature and electrical properties, e.g., electrical conductivity and complex permittivity. We have observed that the DEA properties of the air-aged liquid fatty acids indicate that the electrical conductivity and complex permittivity can be correlated with the degree of unsaturation. It is our objective to establish a relationship between the amount of unsaturation, number of double bond sites and the electrical properties, complex permittivity, and electrical conductivity.

Experimental Protocols and Samples

A TAI 2920 modulated temperature DSC was used in a cool-heat-cool cycle with a ramp rate of $5\,°C/min$ in a nitrogen atmosphere with aluminum pan and lid. Sample size was approximately 5 mg.

A TAI 2970 dielectric thermal analyzer was used to evaluate the dielectric properties in a dry nitrogen atmosphere. The liquid samples were examined with a gold ceramic single surface interdigitated array sensor employing 40 mg of a liquid or solid.

The essential fatty acids were purchased from Aldrich. Oleic acid, a fatty acid, was secured from the College of Pharmacy, University of Toledo, Toledo, OH. The polyunsaturated fatty acids, the essential fatty acids, studied were linoleic (9,12-octadecaidienoic, with two double bonds), linolenic (9,12,15-octadecatrienoic, with three double bonds), and arachidonic (5,8,11,14-eicosatetraenoic, with four double bonds).

The dielectric properties of importance are the permittivity, e', a measure of dipoles and the loss factor, e'', the energy to align dipoles and move ions. The electrical conductivity is determined from a calculation of e'' times the frequency (Hz) times a constant. The complex permittivity, combines a measure of the permittivity and loss factor where $C^{*}=[e'^{2}+e''^{2}]^{1/2}$. Further the tan delta function is the ratio of $[e''/e']$ or the loss factor/permittivity. Since the DEA is an alternating current technique the frequency for the interdigitated array measured in Hz is varied from 0.1 Hz to typically 10 000 Hz. We have observed that the low frequencies of ≤ 1 Hz are specifically related to surface reaction at the gold ceramic interdigitated transducer [2,3].

Results and Discussion

Dithiobis (*N*-succinimidyl propionate) (DTSP) also known as Lomant's reagent adsorbs onto gold surfaces through the disulfide group, so that the terminal succinimidyl groups allow further covalent immobilization of amino-containing organic molecules or enzymes (e.g., horseradish peroxidase, HRP) [4,5]. The reaction of the DEA single surface gold electrode with DTSP at $37\,°C$ was only observed at multiple frequencies from 0.1 to 1.0 Hz. This reaction was used here to define the frequency range where DEA surface reactions occur.

The temperature response of the DEA and DSC systems were tested by examining United States Pharmacopoeia standards in this study. They include the DSC confirmation of vanillin melting at $81–83\,°C$ and acetanilide at $114–116\,°C$, see DEA results for vanillin in Fig. 1. There is a significant response in the DEA and DSC at the melting transition for these two standards at their known melting temperatures where

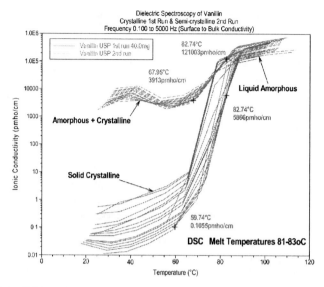

FIG. 1—*DEA calibration at (0.10-5000 Hz) vanillin melt temperature at 82°C.*

the electrical conductivity undergoes a six order of magnitude change from the crystalline vanillin or acetanilide to the amorphous phase.

The oleic acid DSC profile is that of a mixture of components. There are two closely related crystallization peaks and a double melting peak, see Fig. 2. The crystallization occurred at 0.6 to −1.4 with a heat of 6.8 J/g and a repeatable peak at −9.2 to −9.4 (first run) and −9.5°C (second run) with a crystallization heat of 69.2 J/g (total heat 76.0 J/g). The melting peak was −4.1 to 4.7°C with a heat of fusion of 82.9 J/g. The literature melting temperature for oleic acid is 13–14°C [6] and is higher than observed in this study. However, it is obvious that the oleic acid tested here was a mixture with several observed

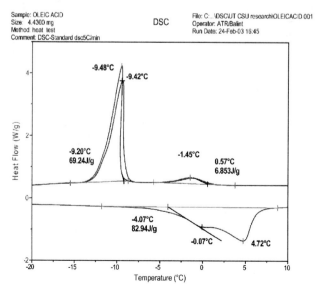

FIG. 2—*DSC cool-heat-cool cyclic DSC of oleic acid.*

TABLE 1—*Melting and crystallization profile of linoleic acid by DSC.*

Cycle	Tc (°C)	Crystallization Tc/p (°C)	ΔHc (J/g)	Tm (°C)	Melting Tm/p (°C)	ΔHf
Cool 1	−19	−17	111			
Heat 1				−8.1	−5.2	118
Cool 2	−18	−16	114			
Heat 2				−8.2	−5.2	117
Average	−19	−17	112	−8.2	−5.2	118
Tc	Crystallization Temperature					
Tc/p	Crystallization Peak Temperature					
ΔHc	Heat of Crystallization					
Tf	Melting Temperature					
Tf/p	Melting Peak Temperature					
ΔHf	Heat of Fusion					

melting temperatures that were lower than the 13–14°C, probably a melting point depression due to impurities.

The DSC curve for linoleic acid in a cool-heat-cool-heat cycle represents twice the crystallization and melting of the acid. This technique establishes the repeatability of the method as follows, see Table 1: crystallization peak temperatures −16°C and heat 114 J/g plus the heat of fusion peak temperature −5.2°C, melt temperature −8.2°C, and heat 117 J/g. The precision is excellent. Crystallization and melting temperatures for the four acids were of the same precision.

The value of the literature melting temperatures for linoleic acid are −5 to −9°C (see Ref [7]) and this study −8°C, as well as the value of the literature melt temperatures for linolenic acid −11 to −17°C (see Ref [7]) versus this study −14°C were both in good agreement. The arachidonic acid DSC melting temperature was −18°C, higher than the reported literature value of −50°C [8]. This enhanced melting temperature was probably due to an arachidonic acid mixture with oxidized species since it rapidly oxidized in air [8].

The electrical conductivity for a first and second run of arachidonic acid in nitrogen at 50 Hz is reported in Fig. 3. The second run retained only 16–22 % of the electrical conductivity or ca. 80 % was lost upon heating from 60 to 140°C and then cooling and reheating. The sample after testing was a hard oxidized resin while the original was a fluid liquid. The effect of temperature on the conductivity rate of

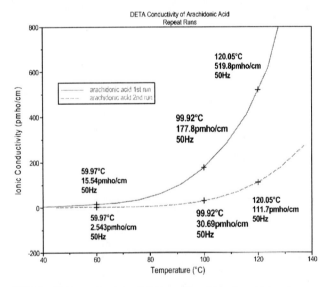

FIG. 3—*DEA conductivity (50 Hz) of arachidonic acid in nitrogen.*

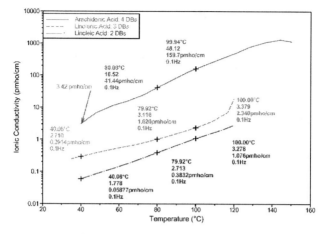

FIG. 4—*DEA conductivity (0.1 Hz) proportional to double bond content in free fatty acids.*

the first and second run determined from Arrhenius plots, log conductivity versus 1/temperature in K were the same. The activation energy was 68 J/mole (first run) and 64 J/mole (second run). The linear equation for this kinetic analysis had essentially the same slopes and intercepts. This implies that the electrical process was most probably the same but at a reduced value on the second run.

Qualitatively the electrical conductivity of the free fatty acids were proportional to the double bond content of the arachidonic acid (four double bonds, DBs), linolenic acid (three DBs), and linoleic acid (two DBs) at 0.10 Hz in the temperature range of 40 to 100°C, see Fig. 4. Linolenic (DBs at 9, 12, 15) acid and linoleic (DBs at 9, 12) acid with similar double bond distributions and markedly different from arachidonic (DBs at 5, 8, 11, 14) acid were clearly delineated in the DEA curves of Fig. 4. There was no apparent pattern of conductivity that mirrored the overall ratio of DBs of 4:3:2 for the acids studied. Arachidonic acid had the highest conductivity for the three acids at all temperatures.

The complex permittivity at 1 Hz versus temperature clearly differentiated the air-aged oxidized linolenic acid and the as-received sample in nitrogen, see Fig. 5. The ratio of complex permittivity of the oxidized acid to nonoxidized (in nitrogen) acid varied with temperature from 294 at 60°C, 982 at 100°C, and 680 at 120°C. The maximum increase due to the oxidation process was ca. 1000-fold at 100°C. The Arrhenius plots for log complex permittivity versus 1/temperature in K rendered activation energy for the oxidized acid of 150 J/mole and 40 J/mole for basic electrical conduction/permittivity process. Normalizing the Ea (oxidation) 150 J/mole for three double bonds is 50 J/mole/DB. The oxidative process had a greater impact on the unsaturation of linolenic acid than the electrical response in nitrogen.

The complex permittivity at 1 Hz versus temperature again clearly differentiated the air-aged oxidized linoleic acid and the as-received sample in nitrogen, see Fig. 6. The ratio of complex permittivity of the oxidized acid to nonoxidized (in nitrogen) acid varied with temperature from 6.2 at 60°C, 12 at 100°C, and 31 at 120°C. The maximum increase due to the oxidation process was ca. ten-fold at 100°C. The Arrhenius plots for log complex permittivity versus 1/temperature in K rendered activation energy for the oxidized acid of 48 J/mole and 20 J/mole for basic electrical conduction/permittivity process. Normalizing the Ea (oxidation) 48 J/mole for two double bonds is 24 J/mole/DB. Again the oxidative process had a greater impact on the unsaturation of linoleic acid than the electrical response in nitrogen.

In summary, the DEA complex permittivity, C^*, at 1 Hz and 100°C ranked the nonoxidized acids as seen in Fig. 7. The complex permittivity was as follows: arachidonic acid 296> linolenic acid 18.8> linoleic acid 4.85> oleic acid 2.6, or C^*: 4 DBs>3 DBs>2 DBs>1 DB at 1.0 Hz (surface properties). The DEA complex permittivity, C^*, ranked the oxidized acids as seen in Fig. 8. The complex permittivity, C^*, was as follows: linolenic acid 4800> linoleic acid 32.2> oleic acid 18.6, or C^*: 3 DBs>2 DBs>1 DB at 1.0 Hz and 100°C (surface properties). The oxidized arachidonic acid was a hard resin and could not be further examined by DEA.

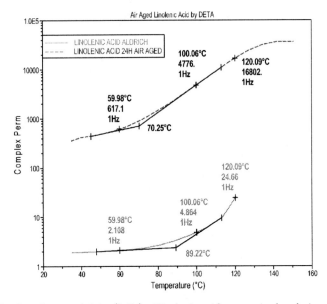

FIG. 5—*Complex permittivity (1 Hz) of linolenic acid: as received and air oxidized.*

Conclusions

The amount of unsaturation from one to four double bonds in the fatty acids was inversely related to the subzero melt temperature in Kelvin.

The as-received acids conductivity at 0.10 Hz was ordered based on the amount of unsaturation and electrical conductivity (PS/cm): 4 DBs > 3 DBs > 2 DBs. DEA complex permittivity, a combination of permittivity and loss factor, also ranked the as-received acids in a dry nitrogen atmosphere at low frequencies (≤1.0 Hz, surface properties). Air oxidation of the free fatty acids was monitored and ordered by

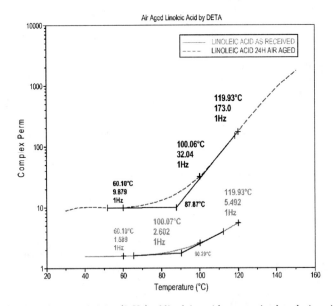

FIG. 6—*Complex permittivity (1 Hz) of linoleic acid: as received and air oxidized.*

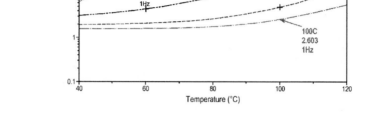

FIG. 7—*Complex permittivity (1 Hz) of arachidonic acid, oleic acid, linolenic acid, and linoleic acid: as received not exposed to air.*

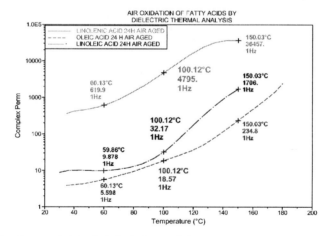

FIG. 8—*Complex permittivity (1 Hz) of oleic acid, linolenic acid, and linoleic acid: air aged.*

DEA at 0.10 to 1.0 Hz and 50 Hz. Dielectric properties followed the oxidation process of the essential fatty acids studied where degree of oxidation as measured by complex permittivity was proportional to the amount of unsaturation in the free fatty acids. We proved our objective and established a qualitative relationship between the amount of unsaturation in the fatty acids and the electrical properties, complex permittivity, and electrical conductivity.

References

[1]   Hitaguchi, H., Ohkubo, K., Ogo, S., and Fukuzumi, S., "Oxidation of Unsaturated Fatty Acids by One Electron Transfer," *J. Phys. Chem.* Vol. 110, 2006, pp. 1717–1725.
[2]   Riga, A., Cahoon, J., and Pilaet, J., "Characterization of Electrorheological Processes by Dielectric Thermal Analysis," *Materials Characterization by Dynamic and Modulated Thermal Analytical*

Techniques, ASTM STP 1402, A. Riga and L. Judovits, Eds., ASTM International West Consho-hocken, PA, 2001, pp. 139–156.

[3] Riga, A., Cahoon, J., and Lvovich, V., "Characterization of Organic Surfactants and Dispersants by Frequency-Dependent Electrochemical and Dielectric Thermal Analysis Techniques," *Materials Characterization by Dynamic and Modulated Thermal Analytical Techniques*, ASTM STP 1402, A. Riga and L. Judovits, Eds., ASTM International West Conshohocken, PA, 2001, pp. 157–173.

[4] Darder, M., Takada, K., Pariente, F., Lorenzo, E., and Abruna, H. D., *Anal. Chem.* Vol. 71, 1999, pp. 5530–5537.

[5] Patolsky, F., Zayats, M., Katz, E., and Willner, I., *Anal. Chem.*, Vol. 71, 1999, pp. 3171–3180.

[6] Sax, N. I., *Dangerous Properties of Industrial Materials*, Vol. 2, Reinhold, NY, 1965, p. 1060 (oleic acid).

[7] Sax, N. I., *Dangerous Properties of Industrial Materials*, Vol. 2, Reinhold, NY, 1965, p. 939 (linoleic and linolenic acids).

[8] Lewis, R. J., *Sax's Dangerous Properties of Industrial Materials*, Vol. 9, Reinhold, NY, 1996, p. 265 (arachidonic acid).